Non-Representational Theory & Health

Non-representational theory is an academic approach that animates the active world; its taking-place. It shows how material, sensory and affective processes combine with conscious thought and agency in the making of everyday life.

This book offers an agenda for health geography, providing the first comprehensive overview of what 'more-than-representational' health geography looks like. It outlines the basis of a new ontological understanding of health, and explores the key qualities of 'movement-space' that are critical to how health emerges within the assemblages that enable it.

It shows how non-representational events and concerns are key to human happiness and wellbeing, to the experience of health and disease, to activities that add to or detract from health and health care work, not to mention to the broader initiatives and operation of health institutions and health sciences.

This book bridges the gap between non-representational theory and health research, and provides the groundwork for future developments in the field. It will be of interest to students, researchers and professionals alike working in health, geography and a range of other disciplines.

Gavin J. Andrews is Professor in the Department of Health, Aging and Society and Associate Member of the School of Geography at McMaster University, Canada.

Geographies of Health
Series Editors

Allison Williams, *Associate Professor, School of Geography and Earth Sciences, McMaster University, Canada*

Susan Elliott, *Professor, Department of Geography and Environmental Management and School of Public Health and Health Systems, University of Waterloo, Canada*

There is growing interest in the geographies of health and a continued interest in what has more traditionally been labeled medical geography. The traditional focus of 'medical geography' on areas such as disease ecology, health service provision and disease mapping (all of which continue to reflect a mainly quantitative approach to inquiry) has evolved to a focus on a broader, theoretically informed epistemology of health geographies in an expanded international reach. As a result, we now find this subdiscipline characterized by a strongly theoretically informed research agenda, embracing a range of methods (quantitative, qualitative, and the integration of the two) of inquiry concerned with questions of: risk; representation and meaning; inequality and power; culture and difference, among others. Health mapping and modeling has simultaneously been strengthened by the technical advances made in multilevel modeling, advanced spatial analytic methods and GIS, while further engaging in questions related to health inequalities, population health and environmental degradation.

This series publishes superior quality research monographs and edited collections representing contemporary applications in the field; this encompasses original research as well as advances in methods, techniques and theories. The *Geographies of Health* series will capture the interest of a broad body of scholars, within the social sciences, the health sciences and beyond.

Also in the series

Forthcoming:

Geographies of Plague Pandemics
The Spatial-Temporal Behavior of Plague to the Modern Day
Mark R. Welford

Published:

Non-Representational Theory & Health
The Health in Life in Space-Time Revealing
Gavin J. Andrews

Non-Representational Theory & Health

The Health in Life in Space-Time Revealing

Gavin J. Andrews

Routledge
Taylor & Francis Group

LONDON AND NEW YORK

First published 2018
by Routledge

2 Park Square, Milton Park, Abingdon, Oxfordshire OX14 4RN
52 Vanderbilt Avenue, New York, NY 10017

Routledge is an imprint of the Taylor & Francis Group, an informa business

First issued in paperback 2020

British Library Cataloguing-in-Publication Data
A catalogue record for this book is available from the British Library

Library of Congress Cataloging-in-Publication Data
Names: Andrews, Gavin J., 1970– author.
Title: Non-representational theory & health : the health in life in space-time revealing / Gavin J. Andrews.
Other titles: Non-representational theory and health
Description: Milton Park, Abingdon ; New York, NY : Routledge, 2018. |
Series: Geographies of health series | Includes bibliographical references and index.
Identifiers: LCCN 2017042158 | ISBN 9781472483102 (hbk) |
ISBN 9781315598468 (ebk)
Subjects: LCSH: Medical geography.
Classification: LCC RA792 .A54 2018 | DDC 614.4/2–dc23
LC record available at https://lccn.loc.gov/2017042158

ISBN: 978-1-4724-8310-2 (hbk)
ISBN: 978-0-367-59263-9 (pbk)

Typeset in Times New Roman
by Out of House Publishing

This book is dedicated to San and Justin who have provided all the love and energy to my life in Canada. I often cringe when I read dedications to family because they so often include the author's apologies for being physically and mentally absent from family life during the production of their product. My typical thought when I read these being, 'well I'm sure they would rather have had you around than endure this involuntary sacrifice and read your dedication'. Hence, I hope to have produced this book without San and Justin noticing too much, time-wise as least (however I do apologize for boring you both with occasional monologues on onflow, affect, space and time etc.).

Contents

Figures and Tables

Figures

Tables

Why now and how? Preface and plan

A professional moment...

'Energy! Energy! Energy!' is the mantra of a recent paper by Chris Philo and colleagues (Philo et al., 2015). On one level they were calling for 'new energy geographies'; for academic attention to different/non-conventional forms of energy, and to thinking about energy in new ways (something that, as we shall see, this book aims to do). On another level, however, with both the lively delivery of the paper itself and in a few of its final words, the authors also dangled the idea of energetic conversation, something more-than-subject matter; that academics and academic objects also create energy. This idea is certainly interesting and deserving of some attention but, in the current context, it evokes two immediate questions. What gave me the energy to write this book dedicated to non-representational theory (NRT) and health? What gives this book, as an object, some energy and energy creating potential?

In terms of the former question, on reflection, I think that there are really a number of motivating factors, or energizing boosts, behind this book. The first is my own experience of health, and how I have come to appreciate particular ways in which it surfaces. I am not an overly spiritual person and nor, although I study holistic medicine from time to time, am I personally invested in alternative explanations for health and wellbeing. That said, I have become increasingly aware that in a world where physical threats loom, economic insecurity and social inequalities persist, work life is long, stressful and uncertain, connections and commitments seem endless, and illness and old age seem imminent – the basics and the moment matter. What as humans we are doing, what we sense, what is taking place around us, and whether or not they give us joy, matters. Indeed, compared to other facets of life, the immediacy of the human experience seems somewhat malleable, positivity here somewhat attainable; a place of hope. Such joy is, of course, not always organic and free-flowing (such as encountering a lovely view whilst out on a walk), and can be encouraged by design; textures and atmospheres engineered by public and private interests to entice and engage us (Thrift, 2008). One only has to, for example, enter a busy restaurant, sports arena, shopping mall, spa or even hospital to experience this first hand. I have become convinced, however, that both organic and artificial experiences beg academic comprehension.

A second motivating factor behind the book is thinking about how we might go about researching these kinds of things. For me, the recent turn to NRT in human geography has provided great inspiration and potential by escaping the social scientist's tendency to automatically stop the world and suck meaning out of it, by instead focusing substantively on the present, on the happening of life. Currently in health geography NRT is new and exciting, yet it is also sporadically employed, slightly incoherent in terms of being a body of work, and is certainly conceptually under-articulated. These issues, I felt, could be partly addressed by a book-length publication that locates ideas, interests, developments and debates in one easily accessible place. Indeed, NRT constitutes an atypical approach that looks at the world differently, and hence a book might help foster a greater academic understanding and appreciation of it. Moreover, a book might help foster a particular mindset and imagination required in NRT; a sense of wonderment with life's immediacy. A fascination with the never-ending stream of less-than-fully consciously acted and experienced existence that, although merely a slither, is where human beings always occupy. It took me a long time to get my head around this part of life and its importance, and how any research approach or tradition could possibly claim to study it. But when you 'get it', when you start to comprehend NRT, a whole new world comes into view. Importantly, it's not just joy and wellbeing that NRT can be concerned with. How the immediate, material and the sensory matter to, and overflow, conscious thought and action in the seemingly mundane worlds of policy making, clinical practice and the delivery of public health are equally important questions.

The third motivating factor behind this book is the current moment in my own career. A few years ago I was in my early 40s, fifteen plus years out from completing my PhD. I was not quite facing a mid-career or mid-life crisis (at least I don't think I was), but I was certainly running out of empirical topics that excited me, and running out of theoretical lenses that inspired me, whilst also realizing that I had over 20 years remaining until my retirement. Having travelled various theoretical terrains – through political economy in the late 1990s, through social constructionism and various post-structuralist theory in the 2000s – in the early 2010s I bumped into NRT. For me it was a revelation, providing a new window onto the world. One which allowed me to get excited about my work by re-looking at the empirical areas I had long considered – holistic medicine, nursing, sport, health histories, ageing etc. – but at different aspects of their existence and my and other peoples' engagements with them. NRT provided another way of doing health geography that did not involve digging for information in order to construct an author narrative (hence departing from much of what happens in space and time and much of what people feel). A particularly important moment in this journey was attending the 2011 International Medical Geography Symposium in Durham, UK. I was sitting with Chris Philo in a pub reminiscing about the Lampeter geography department (where I was an undergraduate and he had started his professional career). He asked me a short and simple question: 'Where's your book?' It was a friendly inquiry that he probably would not even recall now, but one that stuck with me. He was right, despite a diversity of academic output, I had not produced an idea or object that I could proudly point to that had provided some new energy for my sub-discipline. Wil Gesler, for example, had given health geography the idea of therapeutic landscapes (Gesler, 1992), Robin Kearns had nudged medical geography towards a 'reformed post-medical geography of health' (Kearns, 1993), but I had done nothing of the sort. Although I did not start writing a book proposal immediately, it became inevitable from that moment on.

In terms of the latter question – what energy I want this book to have and others to get from it – in short it would be the same fascination with the immediacy of health that I have found, and motivation to pay some attention to it. Indeed, the book's aim is to provide an introduction to NRT for health geography students and researchers breaking into the field and/or curious about its potential. I want the book to entice and encourage further developments by building bridges between NRT and their empirical interests and concerns. Importantly, I am very aware that, at most, I want to articulate and encourage NRT as another string to health geography's bow, not something that should or could replace existing approaches.

With regard to other books that are currently available, a number in human geography do outline, debate and progress NRT in various ways, but they are written for a broad audience and, for the most part, do not specifically deal with health (see Thrift, 1996, 2008; Anderson and Harrison, 2010; Vannini, 2015a). In health geography a few books either foreground certain aspects or interests of NRT or touch upon them briefly, but these are far from comprehensive engagements often made in the context of quite specific empirical considerations (e.g. Foley, 2010). To date, only Cameron Duff's excellent *Assemblages of Health* draws close to this book (Duff, 2014a). Focused specifically on Deleuze's empiricism, *Assemblages of Health* indirectly addresses the 'why' and 'on what basis' of NRT in health contexts. In contrast, and as will become clear, this book directly addresses the 'what has', 'what to' and 'how to' of NRT in health contexts. Hence, the two books are highly complementary and could be absorbed in close order.

A quick spin through the book...

Chapter 1, 'New intellectual energies: the emergence and basis of non-representational theory', sets the disciplinary and theoretical scenes for the book so that readers are well prepared for the arguments that follow in the subsequent seven chapters. As an initiation to NRT, it leads readers to key developments, literatures and debates that have set up and constituted the approach to date. Chapter 1 rounds off by outlining the main critiques of NRT – getting them out of the way early in the book and providing some

early responses – and by summarizing the main empirical fields that constitute current more-than-representational health geography. In terms of definitions, here the term 'more-than-representational' incorporates a range of
scenarios including (i) research that obviously and explicitly is or uses NRT,
and (ii) research that aligns with NRT's principles, and/or uses similar ideas,
and/or looks at least partly at non-representational things in the world, yet
does not explicitly claim to be, or be using, NRT (see Lorimer, 2005). 'Health
geography' meanwhile is also broadly defined as (i) research by all 'brands' of
human geographers – including and beyond self-identifying health geographers – which has a health focus or implication, and (ii) geographical/spatial
research by non-geographers which has a health focus or implication. These
are liberal typologies that hold for the remainder of the book, particularly
where studies in more-than-representational health geography are mined, the
data from them helping to illustrate some key concepts and ideas.

Chapter 2, 'Rethinking health: from what it means to how it becomes',
starts by considering how health and health care have traditionally been
understood and approached in health geography, ranging from focusing on
aspects that are measurable and locatable, to things that are socially constructed and meaningful. It is argued that research currently pays attention to
the spatial contexts that are actively engaged by individuals where and when
health and health care arise. However that, through adopting NRT, research
might also move beyond this and think processually about the composition
and working of assemblages through which health arises and is performed.
Finally, some indications are given of precedent: to how assemblage thinking
has been employed in more-than-representational health geographies to date.

Chapter 3, 'Key conceptualizations in more-than-representational health
geographies', explains some of the core theoretical ideas and concepts in NRT
and traces how they have been used to frame, or at least can be observed
in, more-than-representational health geographies: relational materialities,
onflow, vitality, virtuality and multiplicity, and the hope in potential and
becoming.

Chapter 4, 'A mode of health transmission: affective health geographies',
builds on the previous chapter (affect, being the main working concept of
NRT, deserving dedicated attention in its own right). The chapter begins
by explaining the philosophical origins of affect, particularly in Deleuzian
thinking, and the ways in which it, as a producer of collective energy and
atmosphere, works for and against health. Following this, attention is paid
to the various engagements with affect in more-than-representational health
geographies particularly within the empirical fields of health care, chronic
disease, health-harming behaviours, and the politics and wellness in movement activities.

Chapter 5, 'Characteristic styles and priorities of more-than-representational
health geographies', considers the five substantive concerns of NRT that are
non-representational features of the world, and of the human condition and
experience, that researchers often animate in their empirical studies: practice

and performance, foregrounds and backgrounds, senses and sensations, impulses and habits, and the ordinary and every day. For each, attention is paid to how they arise in emerging more-than-representational health geográphies.

Chapter 6, 'Rethinking movement in health geography: from a change of location to movement-space', deals, as its title suggests, with movement; the basic ingredient and basis of all the concepts and concerns outlined in the previous five chapters. To start, the chapter traces conventional engagements with space, place and movement in health geography, ranging from the macro-scale movement of populations and diseases to the movement – or lack of movement – in people with unhealthy bodies. The argument is developed that whilst the current research engagement with movement is broad, it could be progressed considerably by considering the processual and performed elements of movement, and how they create space and time. Drawing on the work of Merriman and others, the chapter develops the specific idea of 'movement-space', and provides an overview of how 'flow' – a general movement experience and sensation – has arisen in more-than-representational health geographies.

Chapter 7, 'Qualities of movement-spaces in more-than-representational health geographies', explores some core qualities of movement-space (specifically those that make up the character of flows – as introduced in the previous chapter): speed, rhythm, momentum, imminence, encounter, and the immediate contrast of stillness. For each quality a basic scientific explanation is provided, followed by an explanation of how it arises in the social world and in traditional social science research. Attention is then paid to how NRT might instead understand the particular quality, and to how it has thus far arisen in more-than-representational health geographies.

Chapter 8, 'Research practices and future directions', outlines some fundamental methodological priorities and challenges of NRT (including new dispositions and mindsets, and 'witnessing', 'acting into', and 'boosting' the world). It then moves on to review some well-aligned methodological innovations in more-than-representational health geographies, including progressive ethnographies, lively interviews, using new technologies, working with the arts, and new styles and forms of writing and presenting. The chapter finishes by considering future directions for scholarship including thinking about the 'ethics of assemblage': how, as researchers, we might seek to understand health assemblages and what, as agents, we might do to, with and within them.

Readers will notice that most chapters do not include substantive concluding sections. This is a purposeful feature of the book aimed at sustaining its flow and its building of an overall argument.

Acknowledgements

My friends, relatives and colleagues have been part of numerous discussions over the years that have helped lead me to this point. They have together helped produce an intellectual but not pretentious environment where it's good to talk about ideas, be critical and challenge norms. The complete list would be far longer, but these people include Craig Ball, Neil Forrester, Mark Sudwell, Nick Pearson, Dave Rowbury, Justin Andrews, James Andrews, Philip Andrews, Stephen Andrews, Sandra Chen, David Brodie, David Phillips, Nick Jewitt, Josh Evans, Mark Skinner, Neil Hanlon, Mark Rosenberg, Jeff Masuda, Janine Wiles, Cameron Duff, Amanda Grenier, Paul Kingsbury, Robin Kearns, Tim Brown, Andrew Power, Jen Lapum, Elizabeth Peter, Denise Gastaldo, Dave Holmes, Blake Poland, Peter Coyte, Susan Elliott, Ronan Foley, Julia Gray, Rob Wilton, Ed Hall, Jim Dunn, Carol and John Andrews.

Special thanks go to Chris Philo for his early support and advice on what the book could look like and do, to Paul Blythe for producing the excellent cover photograph of Teignmouth, Devon, and to Stephen Andrews and friends for producing the various photographs that are inserted throughout the book. They were given an open opportunity to capture and animate some of the events and feelings conveyed in the book, and the results are excellent.

1 New intellectual energies

The emergence and basis of non-representational theory

We all know the feeling, often vague and in the background yet very familiar. The feeling of not only being present in the moment and participating in the moment, but experiencing life's basic movement and energy. The passion pushing it and us forwards; the surge and the ride. It can be a subtle undercurrent, adding momentum to what we are doing, or a powerful force that sweeps us up and moves us along. It arises through the actions and motions of people and things, but it seems to have a life of its own; something more than the sum of its parts. It is basic, physical and not often consciously reflected on, yet it has a bearing on our immediate sense of wellbeing because being in sync with it, and working in the same direction as it, can be an energising and joyful experience. On the other hand being out of sync with it or moving against it can be difficult, draining and frustrating. It's an experience that happens to greater or lesser extents in all areas of, and places in, our lives; in our homes, workplaces, where we are entertained and spend our leisure time, where we shop and where we are educated and are cared for. It is an important but academically neglected part of the world engaged by non-representational theory.

A group of nurses are at their work station for the last hour of their shift; a geometrically shaped beige island home to computer screens, plastic furniture and wires bathed in artificial light, all but the smallest of corners observable through its see-through acrylic barrier. Despite the nurses' tiredness, many things happen that make it a busy time for them, shortening the hour. Some of these things involve a great deal of their conscious thought and action, particularly regarding the latest institutional infection control strategy which seems to be the order of the day (e.g. attention to the leaflets given to them, patients' families and doctors appearing and asking questions, their shared notes on the day's practices). Some of these things do not require much conscious thought or action on their part (e.g. various internet information that flashes unread on their computer screen, the sun's warmth filtered through a window hitting their faces, bodies moving past in various directions, and background chitter chatter). Meanwhile some of what happens they do not think about at all (e.g. their breathing, repetitive typing, subtle movements of their legs and arms to undertake simple tasks or for comfort). Regardless of their levels of attention, all of what happens in that hour rolling out constitutes their working lives. All of it has some energy and movement to it, and all of it contributes in some way to the whole. It reflects and affects their mood, alters the feel and content of their communications, and turns the course of meaningful events. Missed by much research, yet not by non-representational theory, are these realities. How life happens all the time and everywhere. How it moves on unabated, constantly becoming. How it unfolds as a stream of action and feeling.

I wrote these two vignettes for a graduate lecture I give on non-representational theory (NRT). In my experience, they provide an initial entry point to NRT as an approach which focuses on life's 'ever-breaking wave'; its happening, its taking-place. Beyond this entry point the key task for my graduate students, and this book, is to think more deeply how these types of events are composed and how they unfold as space-time. Moreover, to think about what NRT does and could contribute to the geographical study of health.

Conceived by Nigel Thrift in the mid-1990s (see Thrift, 1996, 1997), and more widely developed and applied by him and other scholars in the last decade (see Thrift, 2008; Anderson and Harrison, 2010; Vannini, 2015a), NRT is a style of thinking and researching based on a plurality of theories that puts the raw performance of the world front and centre in research inquiries (Simpson, 2010). On one level it communicates the many wordless, unreflective, automatic and accidental practices involved, and their expression. On another level it communicates how these are registered and sensed by and affect humans. Thus, NRT moves the focus of academic inquiry onto the physicality and experience of the active world; the bare bones, basic mechanisms, root textures in the rolling out of occasions (Thrift, 2008). The 'non-representational' in NRT has always been the subject of different interpretations and debate, in terms of what it means and whether it is an appropriate descriptor (Lorimer, 2005; Carolan, 2008; Jones, 2008). However, it is generally accepted that there are two levels to non-representation. The first is concerned with events that are both non-representational in their form (based on acts rather than speech), and non-representable in research terms (that are difficult to show); events that constitute what is fundamentally going on in the world. In other words, rather than being concerned with what humans purposefully represent through language and other communication (such as ideas), and the complex and consciously acted phenomena that often follow (such as rules or policies) – events that, for researchers, are eminently representable – NRT is concerned with how humans act and communicate without purposefully representing; the often prior, less-than-fully conscious presentations of life. As Simpson (2015) suggests, this thinking also constitutes a clear rejection of a form of body 'representationalism' embedded in much contemporary social science which assumes meaning to be something formed in the mind that acts as a precondition for physical action (which shows this meaning); NRT's antidote to this thus being its emphasis on less-than-fully conscious practice, embodiment, materiality and the processual.

The second level of non-representation in NRT is concerned with the nature and implication of non(not) representing such events, and what might instead take its place as a research practice. So instead of representing, NRT aims to 'show' and 'animate' the push of the world; to convey it in ways which reflect as much as possible its physicality, energy and intensity (substantive methodological challenges which will be addressed in the final chapter of this book) (see also Thrift, 2000; Dewsbury et al., 2002; Vannini, 2009; 2015a). In short then, although it is an absolute mouthful, it could be said that NRT rejects the automatic drive in much traditional social science to be representational by representing representable representational processes and events, and instead chooses to be non-representational through animating unrepresentable non-representational processes and events.

Importantly, NRT is not an exercise in trivializing or simplifying the immediacies in life. Rather, despite they being omnipresent, as an approach it opens up the complexity of the present, thinking about the different forces that form it. Nor does NRT exist simply to expose some specific and hitherto unexposed non-representational empirical realities of the world. When Thrift and others have challenged scholars to 'go beyond representation' they meant precisely that, not to dispense with it (as some critics have incorrectly assumed). Hence, NRT is an approach that augments traditional forms of geographical research and knowledge by thinking about how humans consciously shape the immediate, physical and sensory in life, and how these things in turn shape the human experience often less-than-fully consciously. In other words, how immediate, physical and sensory non-representational processes (i.e. the forming and performing of life) overflow and interplay with conscious personal, social, political, economic and institutional representational processes (i.e. what more clearly comes into view), and together how they both contribute to power, significance and meaning in life (i.e. what is more clearly produced). Indeed, NRT understands that representations themselves can only ever be very briefly stable due to their composition from ever changing non-representational events, and how representations themselves might be non-representational in the way that they are performative and intially apprehended (Dewsbury et al., 2002). As Thrift (2008:148) suggests, 'meaning shows itself, only in the living'.

Notably, the emergence of NRT has been motivated by range of factors which, whether they be theoretical arguments or more to do with the organization and culture of human geography, have acted as underlying positive currents, pushing and pulling geographical scholarship in its direction and opening up fertile ground for its development. The following two sections summarize the most influential of these.

Pushes towards NRT

I admit that at times I can get frustrated with geography, much of it seems to follow the logic of the corpse, interested in the broken, the static, the already passed... I have therefore tried to develop an approach which can act as a means of providing something livelier. I call this approach non-representational theory ... The stakes are high because what I am attempting to do is to overturn much of the spirit and purpose of the social sciences and humanities... Human life is largely lived in a non-cognitive world... The varieties of stabilities we call representation can only cover so much of the world.

(Thrift, 2004a:81–89)

We murder to dissect. I cannot think of a more apropos descriptor of the academic enterprise which stultifies and deadens the world via the relentless search for explanation, prediction, and meaning. Rendering post-mortems is the stock-in-trade of meddling intellects. This is the malady that is stimulating streams of non-representational thought and 'performative' geography.

(McHugh, 2009:209)

Five critiques of the dominant research paradigms in human geography have emerged, providing a direct and quite forceful push towards NRT as an alternative way of going about research. The first critique is about missing the active world; that social constructionism in particular has suffocated a good portion of what happens in space and time due to its commitment to theory-driven interpretative searches for meaning, and under the orders, structures and processes that are imposed by researchers who employ it (Dewsbury et al., 2002). Indeed, the thinking is that that all too often scholars report on quite specific things that have happened to their subjects and then almost immediately resort to theorizing and interpreting; to peeling off the layers in order to find the mechanisms, consequences and meanings involved (Thrift, 2008). The critique follows that this process has not only led to quite abstract self-serving theoretical debates and the conveyance of a drained and embalmed version of the world (Thrift and Dewsbury, 2000; Dewsbury et al., 2002), it has also left a gaping hole the social scientific coverage of what goes on in the world. As McCormack (2013) explains, much qualitative research is, using a sports analogy, formulaic 'post-game' interrogation and analysis. By this he means that we (researchers) go all out to find 'the player' (e.g. in health contexts a marginalized, disadvantaged or oppressed patient or sufferer) in a break from, or after, 'the game' (e.g. an illness episode or treatment or recovery determined by biology, structures and powers which caused their predicament) and probe to find out what it felt like (i.e. what we already think we know). The player, for their part, then struggles to convey what happened after the fact. Then, feeling slightly dissatisfied, we press further for them to tell us what it all means; all the while their emotions being demonstrable. Getting some of what we were looking for, we then 'send it back to the studio' (i.e. our university office) for in-depth interpretation and analysis (e.g. often filtered through philosophy) (see also Andrews, 2017a). McCormack (2013) notes that this is all based on an underlying assumption that rigour in research can be achieved through triangulated accounts involving interpretative sense making. The NRT critique is that for a discipline that claims to be in tune with people's lives, this process sure feels lifeless, obscure and texty, and – back to sports analogies – largely misses the game (Andrews, 2017a).

The second critique of the dominant paradigms in human geography is about acknowledging the excessiveness of life and the limitations on research this imposes. As Dewsbury et al. (2002) note, even if researchers wanted to attempt to represent life, given all the millions of micro-events/happenings that constitute its taking-place – all of its liveliness and movement – this could never happen. Life is far too excessive to ever be fully understood or theorized, and this excessiveness needs to be actively celebrated by research, rather than ignored or reduced.

The third critique is about the consequence of always and inescapably being in the moving world. In his early writings on NRT, Thrift (1999) argues that most traditional research approaches attempt to distance themselves from the world in order to assess it, when this is not really possible. Instead Thrift argues that researchers occupy a forever shifting position within a forever moving world, their changing relationalities to others things in the world affecting anything they observe or judge. Hence, NRT involves a new kind of realization on the positionality between the researcher and what is researched, that acknowledges constant fluidity and interplay. Indeed, Thrift (1999) comments:

> Non-representational theory arises from the simple (one might say almost commonplace) observation that we cannot extract a representation of the world from the world because we are slap bang in the middle of it.
>
> (1999:296)

The fourth critique is centred on life as ever constructing. This argues that social constructionism deals in fundamental understandings on constructed categories – such as race, health, sex, nation and place – but that these constitute general and premature conclusions that researchers adhere to (Taussig, 1993; Anderson and Harrison, 2010). The theoretical point here is that even if we were to believe that life can reach a constructed state, this necessarily has to come after earlier states of existence that were less constructed and constructing; that were far more fluid, messy, raw and acted (Taussig, 1993; Anderson and Harrison, 2010). These need to be in some way addressed in research, rather than leading empirical inquiry straight into the collection of data on a constructed subject (moreover into human opinions that are always subjective, relational, and often result from personal emotional (over)amplification). In sum, in contrast to social constructionism, NRT does not perceive there to be a constructed world requiring representation. The idea is instead that, in its purest form, the lived world is an ongoing performance; a physical never ending (re)construction (Thrift, 2004a).

The fifth critique is an empirical observation about the form of the emerging twenty-first-century Euro-Americanized world that should not be approached only by conventional research paradigms. It is claimed that, on one level, this is a deconstructing world in which categories such as sexuality, gender, race and class mean less. On another level it is a world in which, somewhat replacing these, consumption and experience have moved to the foreground on an accelerating plane of velocity (Thrift, 2008). Indeed, in this ever faster world new capitalist commodification strategies engineer the root textures and feels of our lives, adding multiple aesthetics, attractions and distractions that play to multiple senses (Thrift, 2008). Thus our possessions, buildings, neighbourhoods, towns and cities ooze synthetic atmospheres. Our bodies move in an increasing range of environments and relate to an increasing range of objects. Our mechanical and technological obsessions give us new forms of knowledge and awareness of the world, and multiple co-existing timescales and levels of consciousness to our existence (Thrift, 2008). Our new affective sensory pastimes compel, drawing us into them (ranging from environmental and political action to fitness and holistic lifestyles). It is thought that these are a range of parts and symptoms of a greater change, what Thrift (2011) articulates as 'lifeworld inc'; a move from a military–industrial complex to a security–entertainment complex which produces an exaggerated humanity. With regard to the latter, he argues that:

> the entertainment sector has grown in size and influence, becoming a pervasive element of the world. From the base of consumer electronics, through the constant innovations in the spatial customization of pleasure found in mass leisure industries like toys or pornography, through branding, gaming and other spatial practices, to the intricate design of experience spaces, entertainment has become a quotidian element of life, found in all of its interstices amongst all age groups.
>
> (2011:12)

These changes all involve a new emphasis on and privileging of movement in society whereby, for example, the appearance of movement and planes of motion become emphasized over traditional anchors (for example narrative on the history of things). As Thrift (2008) notes:

> [T]he basic cardinals of what we regard as space are [sic] shifting... we are increasingly a part of a movement-space which is relative rather than absolute... in which matter or mind, reality has appeared to us as a perpetual becoming.
>
> (2008:102)

Thrift's argument is that traditional social science approaches are not best placed to engage this faster moving sensory world which is more loosely and differently stuck together, and that we need a new lens, something more closely aligned to it to bring it into some kind of focus. As Duff (2014a) suggests, at the very least, many humanists and social constructionists, whilst still committed to meaning and individualism, must be increasingly aware of the illusion of subjectivity and the mediation of the subject and the self by wider affective forces at play in this newly emerging world. He posits that these researchers cannot surely believe in the stabilities they once attributed to the people and places of the world to the same degree?

The pulls of NRT

[NRT involves] attempting to let loose a certain kind of wild conceptuality which is attuned to the moment but always goes beyond it, which always works against cultural gravity, so to speak.

(Thrift, 2008:19)

The focus falls on how life takes shape and gains expression in shared experiences, everyday routines, fleeting encounters, embodied movements, precognitive triggers, practical skills, affective intensities, enduring urges, unexceptional interactions and sensuous dispositions…which escape from the established academic habit of striving to uncover meanings and values that apparently await our discovery, interpretation, judgment and ultimate representation.

(Lorimer, 2005:84)

Two draws of NRT are clear, the first being about the fundamental nature of what NRT is as an approach that makes it both academically necessary and attractive. Indeed, in addition to its critique of traditional forms and paradigms of geographical scholarship, NRT has arisen through its own set of fundamental realizations on what it is to be human. Commenting on these, McCormack (2008) notes that central to NRT are a number of claims: that humans do not always consciously reflect upon external representations when they make sense of the world, that thinking does not necessarily involve the internal manipulation of picturesque images/representations, and that intelligence is a relational process involving a number of actors (human bodies and non-human objects) as active participants. Thus, as Thrift (2008:7) suggests, 'non-representational theory is resolutely anti-biographical and pre-individual. It trades in modes of perception which are not subject-based'.

Evidently then there is an attractive belief basis to NRT, but also adding to its popularity is a high degree of theoretical compatibility between it and other traditions currently on the rise in human geography – together they perhaps constituting a broader ontological turn. As Anderson and Harrison (2010) note, these include a general movement of post-structuralist thought towards more 'lively' theory (such as Deleuzian); the emergence of hybrid geographies that explore nature/culture and human/non-human relations; public geography and its concern for action, intervention and practice; post-humanistic scholarship that de-privileges the subject and explores (new) materialisms; the relational turn with its emphasis on how bodies and objects are networked and perform together, and old subjects being re-approached with a critical emphasis on practice and experience (such as 'new mobilities'). Also proving attractive to scholars is the freedom and experience of doing NRT. Through NRT they have been able to look at and see the world differently and more fundamentally (not unlike as described in the opening narrative in this book's Preface), and escape the typical social science expectation to submerge oneself – perhaps even wallow – in the theoretical grappling and struggling of the research endeavour. Instead they have been able to get into the immediate action, performance and movement of events. This might be regarded as quite a masculinist allure but, as geographer Phil Crang mentioned to me recently in reflecting how popular 'non-rep' had become amongst British Ph.D. students, 'they don't want to do anything else!'

The second draw of NRT is quite practical and concerned with the changing circumstances of the discipline of human geography. Prevailing sectorial features and conditions, particularly in the UK, have made pursuing NRT an attractive and accessible option for both emerging and established scholars. Initially at the institutional scale, the gradual evolution of the work of scholars originally based at Bristol University, and the academic climate of its geography department, were influential: namely that of Nigel Thrift and his Ph.D. students – including Harrison, Dewsbury, McCormack, Wylie – who have gone on to be leading scholars in the discipline (most now producing their own Ph.D.-prepared and NRT-trained scholars). Indeed, their particular brand of human geography has proved irresistible to a new generation of geographers, leading to the dispersion of NRT nationally and internationally (not unlike how new brands of post-structuralist cultural geography emanated from places like the Universities of Lampeter and Cambridge in the late 1980s and early 1990s). Otherwise, a national-scale influence in the UK has been the government-led RAE/REF (Research Assessment Exercise/Research Excellence Framework) audit of scholarship and in particular the credit given in the geography 'Unit of Assessment' to single-authored theoretical papers that convey 'big ideas' (that get frequently cited). Their contribution to the broader impact agenda being largely internal, on geographical scholarship itself. In this context, NRT has become attractive to researchers who are recognized and rewarded by their employers for developing and debating it as a new way of doing human geography (Anderson and Harrison, 2010).

Theoretical lineages in NRT

It is important to note though that non-representational theory is not in fact an actual theory, but something more like a style of thinking which values practice. It is therefore also best thought of in the plural as non-representational theories. In this plurality, theories of post-structuralists, phenomenologists, pragmatists, feminists, and a collection of social theorists, mix in varying concentrations.

(Simpson, 2010:7)

As indicated earlier and in the above quotation, the word 'theory' in NRT is not meant to be read as singular (Lorimer, 2008; Simpson, 2010). NRT does not signify a single theory, or a clearly mapped amalgamation of theories. Rather it describes a number of 'contact zones' with a range of established theoretical traditions and positions. In engaging with these zones NRT is unashamedly selective, adopting a magpie approach of teasing out particular ideas that serve its purposes the best (Simpson, 2010). The fact that, as we shall see, some of these zones appear to be non-complementary reflects the varied positions taken within NRT and its tangled intellectual genealogy. Moreover, it also reflects the fact that NRT is a quickly evolving intellectual project with many thought leaders each adding to a theoretical 'cooking pot' (who may or may not always agree with one another). It is not the place of this book to completely disentangle these threads and their complexities – which has been more than adequately done elsewhere (Thrift, 1999, 2008; Cadman, 2009; Anderson and Harrison, 2010) – but some examples are given in the following paragraphs.

As Vannini (2009) suggests, NRT endeavours to draw together ideas from a variety of social science sub-disciplines and fields (such as material culture studies, performance studies, ecological anthropology, sensory sociology and political science's radical democracy), and at least connect with science disciplines typically quite remote from geographical thinking (such as neurosciences and molecular physics). Hence, although it is a new geographical tradition practiced predominantly by geographers, it is clearly quite trans-disciplinary in terms of its constitution and ideas.

Although NRT reacts strongly to a number of traditional theoretical orientations and ways of viewing the world (as described in a previous section) scholars do look back and acknowledge a debt to certain arguments in them that they regard as foundational. So for example, although much of humanism's emphasis on the richness and primacy of human narrative does not sit well with the NRT's relational materialism – even to the extent, as Duff (2014a) suggests, that some scholars of NRT think of humanism as promoting 'cultural fiction' – valuable points of its departure for NRT are noted. As Lorimer (2005) observes, Tuan (2004) stresses how belonging is formulated by direct physical experience and immediacy in place (i.e. through how place is repeatedly heard, smelled and touched as an intimate world), and how belonging is re-evoked by recollections of this immediacy. Similarly, although social constructionism is heavily critiqued by NRT in the ways described earlier, scholars acknowledge how both traditions share the same recognition that orders and symbolic orders in life are arbitrary and invented (rather than 'natural'), and can be contested. Moreover, they acknowledge that, although they have very different contact points, both NRT and social constructionism recognize that representation matters (see Anderson and Harrison, 2010). Finally, in the same vein, although certain post-structuralist preoccupations with finding meaning – and related concepts such as construct and self – are largely jettisoned by NRT, other post-structuralist ideas are more readily adopted, such as, for example, Derrida's ideas on materiality, force encounters and relations between things, and Deleuze and Guattari's unorthodox postmodern thinking on desire and machine-produced reality (in these specific cases to help nurture a deeper appreciation of the productive and disrupting capacity of the material world; Cadman, 2009). Indeed, for many, NRT is itself broadly post-structural to the extent that it reacts against structuralist ideas on concrete social realities.

At its core, NRT involves the re-appreciation and/or re-interpretation of five established theoretical/philosophical traditions, the first of these being process philosophy. This tradition has origins in the processual thinking of Whitehead and others that reacts to forms of scholarship that see dynamic features of life as ascetics or consider events as discrete, static, complete and individual. Processual thinking in contrast involves a fundamental recognition that the world is made of ongoing never ending processes; of series of relational actions that follow, and flow, with some logic. In NRT there are three primary engagements with the processual. First, with what happens: 'the processual *of* life' (the action and register of life's process – its moving forward from one thing to another without ever being in a fixed or complete state). Second, with how it happens: 'the processual *in* processual life' (the processes in this physicality and its register). Third, with how to investigate what happens: as a researcher thinking and acting processually to find and describe these processes. Hence, in short, NRT both attends to processual space-time and is itself processual (see Merriman, 2012). In addition to its basis in process philosophy, processual thinking has practical precedent in archeology, which in the 1960s moved beyond categorizing objects to think about how humans create objects and why they do things with them (see Clarke, 1973). The connection here with NRT is not only the shared interest in the ordering of human material practices, but also a realization that, as scholars, we can learn about culture and society from material processes and traces.

The second of these theoretical/philosophical movements is phenomenology. The forefather Heidegger's own work has been re-approached in NRT but, rather than focusing on human meanings and identities obtained from 'being-in-the-world', focusing more on the consequences of humans being unescapably 'thrown-into-the-world' and inseparable from its physicality (thus helping foster a realization that how humans co-create the physical world might be even more important than how they might subjectively reason it; Cadman, 2009). Indeed, as Wylie (2005) points out, Deleuze (1993) in his book *The Fold* critiques Heidegger's ideas, arguing that being-of-the-world always precedes being-in-the-world because the self and the world overlap in an elastic and continuous unfolding. Elsewhere,

Merleau-Ponty's phenomenology of perception has been re-approached in NRT for ideas on the body and its physical experience being the basis for human experience, and more specifically for ideas on pre-consciousness and the co-evolution of bodies and things (Merleau-Ponty, 1962). Otherwise – and as explored in Chapter 3 – Pred and Whitehead's work on onflow has been re-visited for its phenomenological account of the stream of physical and conscious becoming (Pred, 2005). Finally with regard to phenomenology, scholars have recognized that movement and energy have not been regarded in this tradition in the light that they might have, particularly given early ideas that position them as fundamental. Spinney (2015), for example, argues that,

> For Husserl, the perception of movement meant the perception of continuously changing objects and the relations between them – what Husserl called 'relational moments' where we uncover different facets of things by moving round them. For Husserl, movement gives unity or completeness to things because it relates them with possible trajectories in relation to other things. As such, events and relations come to be seen in wider contexts and possess properties not evident if viewed statically. Indeed, according to Husserl, static perceptual content is an abstraction from dynamic content. Hence, there is a phenomenological precedent for wanting to understand in detail the unfolding of movement and how it orients us to and affects our experience of environments.
>
> (2015:234)

Hence, Spinney posits that 'post-phenomenology' is a theoretical tradition that attempts to move beyond phenomenology's reputation as an anti-science that believes that only the self exists, and seeks instead to investigate the critical roles of technologies and non-humans in life. Whatever the merits of 'post-phenomenology' as a title – because of its significant departures – at the end of the day it is well aligned with phenomenology's first and fundamental tenet: that we, as researchers, should be aware of what humans are not fully conscious of yet do (whether this involves basic movements or complex interactions).

The third theoretical and philosophical tradition that NRT draws on is pragmatism. The classical/naturalistic pragmatism of Dewey, James and others has been re-approached in order to think about how thought and knowledge are never something on their own and are only ever a product of the body's adaptation to its immediate environment (i.e. that corporeal routines and rituals produce meaning in embodied movement; Jones, 2008; Vannini, 2009). Indeed, scholars acknowledge that both NRT and pragmatism share core beliefs; that ideologies, ideas and theories are (i) human creations – their character being deeply dependent on bodily contexts and vantage points, (ii) only ever temporary hypotheses that need to be continually re-thought, (iii) only work if they have practical applications and consequences and are observable. Hence both NRT and pragmatism encourage a 'reformed empiricism' (see Jones, 2008) that (i) has an incredulity towards finding truth through reason and representation alone, (ii) acknowledges that researchers cannot conduct research as disinterested/passive/apart from the world, (iii) believes that, during research, researchers must experiment with the world (i.e. manipulate and change it – three final facets/ideas that come through strongly in NRT methodologies explored in detail in Chapter 8) (for more general applications to geography and health see Cutchin, 1999, 2003, 2004, 2008).

The fourth tradition is vitalism (also explored in Chapter 3 and 4). The philosophy of Spinoza, Deleuze and others has been re-approached in NRT to help focus on the liveliness and energies – the vital forces – of the world (Philo, 2007; Cadman, 2009; Greenhough, 2010). Specifically, on vital living bodies, which possess essential spark and energy and capacity to constantly move and evolve (as conveyed by traditional vitalist/neo-vitalist philosophy), doing things with vibrant non-living objects, which possess their own capacity to act as quasi-agents with their own tendencies, trajectories and forces (arising in the more contemporary notion of 'vital materialism'; Bennett, 2009). This entire process is aligned closely with the Deleuzian idea of vital topography, where the 'inside' (nature – such as the mind and consciousness) and 'outside' (material culture – including habits, practices with resources and technologies) are recognized as non-separable, constantly folding into each other (this process being where human life is created and recreated; Deleuze, 1988, 1994; Duff, 2014a). Two vitalistic ideas have been particularly instructive in NRT, one of these being 'immanence'; a commitment to, and belief in, a different way of explaining the origins and manifestation of life. Life not as the result of the structural impositions (such as codes or laws) of external/higher transcendental phenomena (such as god, reason, humans or nature), but life as an autopoietic self-organizing process encompassing and manifested in the actual material world itself (Duff, 2014a). Life being the drawing of matter together in its own field of becoming or 'plane of immanence', where everything (bodies and objects) is on the same level of existence (being only different modes of the same substance), and where repeated modifications and adjustments express as attributes. The other of these ideas – by comparison more immediate and practical – is 'affect', drawn on to describe a transitioning of the body and the process whereby it is affected by other bodies, modifies and affects further bodies, encompassing and creating positive changes in energy and atmosphere. Indeed, affect is a central meta-concept and the main testing ground of NRT, returned to in much greater detail in Chapter 4 (also see Thrift, 2004b, 2008; Anderson, 2006).

The fifth tradition drawn on by NRT is actor-network theory (ANT). Latour's seminal work here has been re-approached to help explain the active role of technologies and objects in agency with humans. NRT gains from ANT a radical view of the world not as an external environment, but rather something that is enacted into being through heterogeneous networks of human and non-human entities. Moreover, it gains from ANT a presupposition of a type of radical relationality between human and non-human entities; they are always understood to mutually constitute each other through associations (as with immanence, both human and non-human being the same, ontologically speaking, and both capable of agency depending upon how they are associated within a network). However, at the same time, NRT departs from ANT in certain ways, for example by not being directly concerned with ANT's veritable toolbox of concepts and processes and explaining these in empirical contexts (such as intermediaries, mediators, translation, convergence, enrollment and so on). Moreover, by not sharing ANT's vision of a symmetrical placeless world of equal human and non-human technical inputs; NRT instead emphasizing the expressive skills and practices of human bodies and the importance of place (Waitt and Cook, 2007). Hence one might argue that NRT provides a less reductionist, more spatial and ultimately more satisfying analysis than ANT.

These five traditions are certainly not exhaustive, scholars also finding precedent and guidance in social ecology, speculative realism, neo-materialism, symbolic interactionism, and also the work of Mead, Goffman, Bergson, de Certeau, Foucault, Derrida, Bourdieu, Haraway, Massumi and others – indeed highly varied scholarship originating from both Europe and North America and clearly responsible for numerous developments in social scientific inquiry for over a century (see McCormack, 2007; Jones, 2008; Cadman, 2009; Vannini, 2009). Calming this diverse engagement with theory, however, is the fact that scholars using NRT in empirical research tend not to get too engrossed in theory for its own sake. In other words, they use theory in their studies of the active world, but do not often get involved in re-informing and re-stating the nature of theory itself; hence theory being a 'modest supplement' to the life that is animated by NRT in new ways (Anderson, 2009). Otherwise, if this was not the case, NRT would fall into the very trap that it attempts to escape. Moreover, as Lorimer (2015) reflects, to keep research lively, he and other scholars must recognize and inhabit the tensions between (i) 'big theory' imported to explain place, (ii) ideas and theory based in the specific and local, and (iii) writing place straightforwardly as one sees and feels it (the natural inclination and urge of the NRT researcher). Hence, Lorimer muses that he 'holds back theory… steadily letting it in'. Ultimately then, theory in NRT is secondary to empirical knowledge, so that telling life does not become displaced in a process of understanding and knowing life (Vannini, 2015b).

More-than-representational health geographies

NRT has gradually filtered its way down and into human geography's various constituent sub-disciplines and substantive fields of inquiry, and health geography is no exception. This is particularly the case if, as suggested in the Preface, one broadens one's perspective and considers 'more-than-representational health geography' – 'more-than-representational' (to reiterate) incorporating a range of scenarios including research that obviously and explicitly is or uses NRT, and research that aligns with its principles and/or uses similar ideas and/or looks at least in part at non-representational things in the world. 'Health geography' meanwhile broadly defined as research by all brands of human geographers which has a health focus or implication, and also geographical/spatial research by non-geographers which has a health focus or implication. Indeed, as Lorimer (2008) notes, NRT is often a 'background hum' rather than an explicit framework in research, informing and tweaking the style, form and techniques of studies, and in no case is this more so than in health geography, where various components and facets of NRT are now being quietly, yet substantially, road tested in health contexts (particularly by scholars such as Cameron Duff, Ronan Foley, Rachel Colls, Ed Hall, Rob Wilton, Josh Evans, Chris Philo, Sarah Atkinson, David Conradson, David Bissell, Candice Boyd and Beth Greenhough). These studies will be returned to and mined extensively in later chapters, but for now a brief overview of empirical and other themes is helpful.

Empirically in health geography there has been a focus on the feelings and performances – and networking of bodies and objects that create these – in health care and public health, including in holistic practice (Andrews, 2004; Doel and Segrott, 2004; Paterson, 2005; Ducey, 2007, 2010; Andrews and Shaw, 2010; Andrews et al., 2013; Justesen et al., 2014; Lea et al., 2015) and the conventional contexts and practices of biomedicine, bioscience and biopolitics (Brawn, 2007; Greenhough, 2006, 2011a, 2011b; Andrews and Shaw, 2010; Evans, 2010; Solomon, 2011; Greenhough and Roe, 2011; Anderson, 2012; Andrews et al., 2013; Roe and Greenhough, 2014). A range of community-based therapeutic and/or enabling situations and performances have also been considered for the same components and qualities (see Conradson, 2005; Colls, 2007; Kraftl and Horton, 2007; Duff, 2009, 2011, 2012, 2014a, 2016; Tucker, 2010a, 2010b; Foley, 2011, 2014; Lorimer, 2012; Pitt, 2014; Philo et al., 2015; Lea et al., 2015; Barnfield, 2016) – although health-harming agency has also been explored (Jayne et al., 2010; Tan, 2012; Ravn and Duff, 2015; Duff, 2014a, 2014b; Duff and Moore, 2015) – including notably in relation to disability (MacPherson 2009a; Stephens et al., 2015; Hall and Wilton, 2017), arts and their practices (Anderson, 2002, 2006; McCormack, 2003, 2013; Evans et al., 2009; Atkinson and Rubidge, 2013; Atkinson and Scott, 2015; Andrews, 2014b, 2014c; Evans, 2014; Simpson, 2014; Andrews and Drass, 2016), and specifically through walking (Macpherson, 2008; 2009a,b; Andrews et al., 2012; Gatrell, 2013). Although their health implications are often implicit, or their focus is primarily on wellbeing and enjoyment, a range of inquiries into sports/fitness movement-activities have also emerged (Spinney, 2006; Waitt and Cook, 2007; Saville, 2008; Evers, 2009; Humberstone, 2011; Green, 2011; Thorpe and Rinehart, 2010; Anderson, 2014, 2014; Lorimer, 2012; Barratt, 2011, 2012; Bissell, 2013; Cook et al., 2016; Foley, 2015; Barnfield, 2016; Brown, 2017; Cook and Edensor, 2017).

Table 1.1 Empirical areas of health geography informed by, or aligned with, or explicitly NRT

Empirical areas of interest	Authors/papers
Care and public health (including holistic practice)	Andrews, 2004, 2011; Doel and Segrott, 2004; Paterson, 2005; Ducey, 2007, 2010; Andrews and Shaw, 2010; Andrews et al., 2013; Justesen et al., 2014; Lea et al., 2015
Biomedicine, bioscience and biopolitics	Hall, 2003, 2004; Brawn, 2007; Greenhough, 2006, 2011a, 2011b; Andrews and Shaw, 2010; Evans, 2010; Solomon, 2011; Greenhough and Roe, 2011; Anderson, 2012; Roe and Greenhough, 2014
Community-based living, enabling and therapeutic situations	Andrews et al., 2014; Conradson, 2005, 2007, 2011; Kraftl and Horton, 2007; Colls, 2007; Duff, 2009, 2011, 2012, 2016; Foley, 2011, 2014; Lorimer, 2012; Barnfield, 2016; Philo et al., 2015; Lea et al., 2015; Tucker, 2010a, 2010b; 2017; Duff, 2014; Pitt, 2014; McHugh, 2009; Wylie, 2009
Disability	Macpherson, 2008, 2009a, 2009b; Stephens et al., 2015; Hall and Wilton, 2017
Health, music and the arts	Anderson, 2002, 2006; McCormack, 2003, 2013; Evans et al., 2009; Atkinson and Rubidge, 2013; Atkinson and Scott, 2015; Andrews, 2014b, 2014c; Evans, 2014; Simpson, 2014; Andrews and Drass, 2016; Boyd, 2016; 2017; Skinner and Masuda, 2014
Walking	Macpherson, 2008, 2009a, 2009b; Andrews et al., 2012; Gatrell, 2013; Wylie, 2005; Middleton, 2009, 2010, 2011
Other movement activities (including 'healthy lifestyles', sports and fitness)	Anderson, 2014; Barnfield, 2016; Barratt, 2011, 2012; Bissell, 2013; Cook et al., 2016; Cook and Edensor, 2017; Edensor and Richards, 2007; Evers, 2009; Foley, 2015; Green, 2011; Humberstone, 2011; Lorimer, 2012; Spinney, 2006; Thorpe and Rinehart, 2010; Brown, 2017; Simpson, 2017
Health-harming activities	Jayne et al., 2010; Tan, 2012; Ravn and Duff, 2015; Duff and Moore, 2015; Duff, 2012, 2014a, 2014b; Bøhling, 2014; Green, 2011

Theoretically in health geography there has been a focus on specific concepts and approaches of NRT. Notably, ANT and the idea of affect have been drawn on; the former to expose the equal importance, networking and active agency of humans and technologies in health and health care through different scales (Hall, 2004; Greenhough, 2006, 2011a; Timmons et al., 2010; Duff, 2011; Andrews et al., 2013; Ravn and Duff, 2015), the latter, using a Spinozian/Deleuzian definition, to illuminate the energizing passions associated with health and care that are trans-humanly created, transported and experienced (McCormack, 2003; Conradson, 2005; Kraftl and Horton, 2007; Bissell, 2008, 2009, 2010; Duff, 2009, 2010, 2011, 2012, 2014a, 2016; Foley, 2011, 2014, 2015; Boyer, 2012; Andrews et al., 2013; Andrews, 2014c; Atkinson and Scott, 2015; Duff and Moore, 2015).

The latest and perhaps the most significant theoretical contributions to this emerging health geography literature are two-fold. On the one hand there has been some detailed consideration of the philosophical foundation for NRT-based health geographies that exists in the work of eminent scholars, such as in Deleuze's transcendental empiricism (Duff, 2010, 2014a, 2014b; Stephens et al., 2015), Canguilhem's vitalism (Philo, 2007) and Foucault's College de France lecture series (Philo, 2012). On the other hand, a number of focused discussion papers have debated, perhaps more pragmatically, the potential upside and downside of employing NRT in health geography as a new direction (see Andrews et al., 2014; Kearns, 2014; Hanlon, 2014; Andrews, 2014a; 2015; Andrews and Grenier, 2015; Hall and Wilton, 2017). Specifically, Andrews et al. (2014) made the case for a more focused and comprehensive investment in NRT in the sub-discipline by illustrating, through an empirical case study, how wellbeing emerges affectively in every-day moments (rethinking wellbeing from a materialist, post-human perspective as emerging as the environment). This paper was in the style of Kearns' famous 1993 contribution that, a generation before and using a similar case study approach, had made the initial pitch for a humanistic 'place-sensitive' geography of health (Kearns, 1993). Kearns (2014) himself responded to Andrews et al. generally positively but highlighted concerns with how research might ever capture events prior to their becoming meaningful, and the potential scope of NRT beyond the case study Andrews et al. used. Kearns' response was also joined by Hanlon (2014) who usefully debated the connections between materiality, feelings and affects in health, drawing attention to particular literatures he thought Andrews et al. had missed. Andrews (2015) replied to these responses, offering some ideas on getting at the non-representable in health, and on combining the non-representable and representable in research.

In sum, as we shall see in the forthcoming chapters, NRT constitutes a break from the traditional 'twin streams' of health geography (mapping health and health care across space vs digging for the meanings of health in place), and tells us instead something new about the networked, emerging, physical, sensory, atmospheric, energetic, performed and moving nature of health and place. In making this break, as health geographer Josh Evans recently observed at a session at the 2017 International Symposium in Medical Geography, NRT also constitutes a key component of an 'ontological turn' in health geography. This being a move away from theory as epistemology (the way we know things) to theory as ontology (what things are). Thus NRT involves thinking in new radical ways about – and even fundamentally rethinking – health and place. This book examines the ideas at the heart of this turn.

Table 1.2 Theoretical debates on health geography and NRT

Theoretical uses, developments and debates	Authors/papers
Key thinkers and theory	Philo, 2007, 2012; Duff, 2010, 2014a, 2014b; Stephens et al., 2015; Greenhough, 2011a
Progress in health geography and related disciplines	Andrews et al., 2014; Andrews, 2014a, 2015; Andrews and Grenier, 2015; Kearns, 2014; Hanlon, 2014; Hall and Wilton, 2017; Andrews et al., 2013; Andrews, 2017a, 2017b, 2017c
Actor-network theory	Andrews et al., 2013; Duff, 2011, 2012; Greenhough, 2006, 2011a; Hall, 2004; Timmons et al., 2010; Barratt, 2011, 2012
Affect	McCormack, 2003; Conradson, 2005, 2011; Kraftl and Horton, 2007; Bissell, 2008, 2009, 2010; Duff, 2009, 2010, 2011, 2012, 2014b, 2016; Foley, 2011, 2014, 2015; Boyer, 2012; Andrews et al., 2013, 2014; Andrews, 2014c; Atkinson and Scott, 2015; Duff and Moore, 2015; Evans, 2010; Solomon, 2011; Wylie, 2009; Simpson, 2017
Innovations in methods and practice	Masuda et al., 2010; Roe and Greenhough, 2014; Greenhough, 2011b; Justesen et al., 2014; Dean, 2016; Richmond, 2016; Bell et al., 2015; Anderson, 2014; Wylie, 2005; Boyd, 2017; Andrews and Drass, 2016; McCormack, 2003; Spinney, 2015; Skinner and Masuda, 2014

Building bridges

Because NRT takes some scholars far from their comfort zones and challenges the approaches they have heavily invested in, it does have its detractors in human geography. Specifically there are those who (see Creswell, 2012): (i) take issue with the style in which NRT is presented, particularly what they regard as the evangelical tone of its 'manifesto' and excessive quotation of particular theoretical forefathers; (ii) find the notion of non-representation to be problematic, their argument being that most events in life will eventually be represented and that researchers always have to represent to some extent; (iii) argue that the neglects and deadening effect of social constructionism and other paradigms have been overstated in NRT, pointing to a lack of acknowledgement of how auto-ethnographers, sensory ethnographers and others have undertaken work closely aligned to NRT for many years; (iv) point out that many geographers do try and figure out how representation works in the world, and are well aware and critical of the politics, biases and consequences of their own representations; (v) note the bewildering array of theory and philosophies NRT draws on, each partially and loosely – questioning whether its centre can hold and whether NRT loses touch with anything concrete (thus, if it has the ability to be anything concrete); (vi) point to the irony that, contrary to its stated objectives, much NRT does in fact involve exclusionary languages, excessive theorization and a lack of empirics, it being more about other writers and texts than about the world beyond these texts. Moreover, there are those who (vii) take issue with what they see as the universalist nature of NRT, it being, for them, masculinist, technocratic, abstract and relational to the point of meaninglessness (Thien, 2005; Bondi, 2005; Tolia-Kelly, 2006). For them, NRT distances conscious feelings and emotions (Thien, 2005), fails to differentiate bodies and recognize persons through important social and demographic categories – such as gender, ethnicity, disability and age – and the particular challenges and injustices these groups face (Bondi, 2005; Creswell, 2012), and moreover fails to recognize political power and intent (Jacobs and Nash, 2003; Pain, 2006; Creswell, 2012).

Some of the scholars voicing these concerns probably have entrenched opinions on NRT that will not change. Others might, however, be skeptical of NRT yet still open to persuasion one way or another (whilst others, of course might not yet know enough about NRT to have formed opinions). The remainder of book does not address the aforementioned concerns systematically. It is instead more exemplary in terms of its response; the hope being that seeing NRT applied to health and health care in detail might foster a stronger appreciation of its value and strengths, how it might be integrated with existing approaches and how its limitations might be negotiated.

2 Rethinking health

From what it means to how it becomes

An animal, a thing, is never separable from its relations with the world.
(Deleuze, 1988:125, cited by Duff, 2010)

The old adage used to support historical reflection is that 'one only knows where one's going if one knows where one's been.' There is of course some truth in this saying, and if NRT is to be more fully embraced in health geography, consideration needs to be given to what specifically it might add in terms of understandings and approaches. With this in mind, the first section of this chapter examines the past treatment of health across medical and health geography, including associated concerns (such as for disease, care and wellbeing). Based on the work of Gilles Deleuze, Cameron Duff and others, the second section then challenges this sub-disciplinary engagement by providing a different view of health and how it becomes as part of particular assemblages, and how it might be studied through NRT in a way akin to an ethology. In support, examples are then provided as to how assemblage theory has to date framed and informed empirical studies in 'more-than-representational health geographies' (as defined in the Preface and Chapter 1).

Traditional perspectives on health

States of ill-health: spatial expressions

As is well known, the sub-disciplinary title of 'health geography' is the one favoured by scholars today, it incorporating a broad range of contemporary approaches and interests. Twenty-five years ago, however, 'medical geography' was favoured and better reflected scholarship of the time. This said, as Mayer (2010) reflects, the term medical geography has always been a little misleading because most research was not, and is still not, in fact interested in the same things as medicine (e.g. physiology, diagnosis and treatment), but was and still is far more akin to 'epidemiologic geography' or 'landscape epidemiology'. In other words, it is interested in forms of non-human life (e.g. pathogens, viruses, bacteria and vectors) but, rather than being focused directly on them (i.e. their nature and the way they work), it is focused 'downstream' on their cumulative effects on human life (predominantly in broad population terms). With regard to approaches, this is reflected, for example, in the longstanding adoption of disease ecology which argues that disease must be understood in terms of the contexts within which it exists; attributing spatial patterns of disease to features of populations (such as genetic and demographic attributes), environments (relevant geographically definable factors) and human behaviour (manifest in choices, actions and interactions), which all act as stimuli (May, 1959; Oppong and Harold, 2010). This understanding has led, for example, to the adoption of sophisticated modelling techniques to explain variations in disease between areas by testing different factors and attributes associated with its occurrence (Duncan et al., 1996, 1998; Gould, 2010).

Whatever the approach, however, in medical geography health is simplistically and implicitly thought to be a state whereby disease is absent. An important state to reach, yet empty; not something with character in itself, and thus not something that should be focused upon in itself (Kearns, 1993). Indeed, there is no real perspective on health per se in medical geography, just on the spatial expression of things that detract from it. For example, a strong focus has developed in recent years on 'the social determinants of health' and 'place-effects on health' (as it has in many other disciplines across the health sciences). For a substantial period in the 1990s and into the new millennium, debates have persisted across the sub-discipline as to whether health is more greatly influenced by the demographic and other characteristics of local populations (social composition), or by the services, facilities and other infrastructure available to them (social contexts) or by their cultural beliefs and actions (collective social agency) (Macintyre et al., 1993; Duncan et al., 1998; Ecob and Macintyre, 2000; Macintyre et al., 2002) (notably the latter interest seguing into critical attention specifically to the role and forms of social capital; Mohan et al., 2005; Veenstra et al., 2005). Despite all this attention, however, it might be suggested that even here, the promise of an engagement with health has not materialized, most research being concerned with less-than-perfect health status (e.g. Dunn and Dyck, 2000; Diez-Roux, 2001; Macintyre et al., 2002). Hence, perhaps, given the nature of most studies, the terms 'social determinants of illness', or 'place-effects on illness' might be more appropriate descriptors even for this relatively progressive body of work?

States of ill-health: the body, meaning and experience

Just as medical geography was not really concerned with medicine, one might equally argue that a great deal of health geography is not really concerned with health. This is because a fair proportion of contemporary scholarship is still concerned with states of ill-health, albeit coming to grips far more with the meaning of ill-health, often using qualitative methods. The development of this tradition can be traced to Kearns' argument for a 'post-medical' geography of health in the early 1990s (Kearns, 1993) and later arguments on how to take forward this agenda through an emphasis on 'the body' (Dorn and Laws, 1994; Hall, 2000). At the time this argument very much mirrored a wider emphasis on the body emerging across a number of social sciences (Kelly and Field, 1996) which was critical of medicine's construction and treatment of bodies as mechanical and/or as vectors and/or as transcendable, and instead aimed to take into account phenomenological experience of the body; relevant histories, cultures and politics that impact the body, and its representation (Parr, 2002). In short, it was a perspective that regarded the body as a 'social body' that individuals inhabit their worlds with and hence that should be a complex site of comprehensive study in itself (Parr, 2002). Notably, although this emerging scholarship in health geography has been underpinned predominantly by humanistic and social constructionist theoretical approaches, following Hall's (2000) call for attention to the experienced materiality of the body (the body's 'blood, brains and bones'), shades of 'new materialist' perspective have also played a small part in the most recent inquiries (e.g. Colls, 2007).

One substantive empirical theme of this body-focused health geography has been living with disease. As Del Casino Jr (2010) notes, diseases are not static phenomena but have different impacts across time and space on individuals who mitigate them in different ways under different circumstances. Indeed, Del Casino Jr argues that diseases often fluctuate and are uncertain, and sufferers are often closed off from, or frequently have to physically negotiate and re-negotiate relationships with, a range of places including homes, neighbourhoods, workplaces and clinics – each of these potentially involving the playing out of specific renditions of their illness (Moss and Dyck, 2003; Del Casino Jr, 2010). Specific diseases of interest to geographers have included chronic conditions such as rheumatic illness (Crooks and Chouinard, 2006), MS (Dyck, 1995), ME (Moss and Dyck, 1999; MacKian, 2000) and age-related impairments and care (Wiles, 2011; Antoninetti and Garrett, 2012) (notably the label 'chronic illness' only being made meaningful because of different impacts across time and space and their mitigation; Moss and Dyck, 2003; Del Casino Jr, 2010). Meanwhile, states of mental ill-health have been focused on, ranging from serious diagnoses and conditions (Parr, 1997, 2000) to common phobias and disorders (Segrott and Doel, 2004; Andrews, 2007; Davidson, 2000; 2005). With regard to the former, the emphasis has gradually moved from 'landscapes of despair' to one that, whilst still acknowledging challenges, lays emphasis on pathways to empowerment, enablement and recovery (Parr, 2008; Parr and Davidson, 2010) and community integration (Yanos, 2007; Townley et al., 2009).

Throughout much of this body of work, the ways and processes through which ill bodies are othered and marginalized is a major concern. This is often based in the fact that sufferers may have bodies that look different or have different capacities than the majority of 'normal' bodies they interact with. Either this or their conditions or symptoms may not be invisible and outwardly manifesting, their failing to 'look sick' leading to its own specific challenges related to acceptance and recognition (Moss and Dyck, 2003). Either way, the result is that sufferers may well have to negotiate being judged and treated in particular ways including by health institutions and media (Craddock, 1999; Craddock and Brown, 2010). Finally, nudging scholarship beyond a focus on ill bodies is research on non-ill bodies and their particular challenges in health contexts, such as those labelled as fat (Colls, 2007, 2012a), those pregnant (Longhurst, 2000, 2005) those breastfeeding (Mahon-Daly and Andrews, 2002; Boyer, 2011) and those that work out (Andrews et al., 2005; Nash, 2012).

The avoidance of ill-health/production of health: supports and critiques of public health

One way in which health geographers have engaged with ill-health is to think about how it might be caused or avoided; their arguments supporting or critiquing discourses and practices employed under the new public health movement which has emerged since the 1980s around a broadened under-standing of health (this shifting away from a focus on biomedical models and embracing a broader ecological approach that takes into consideration social, psychological and geographical determinants of health, and employs social education to promote health; Brown and Duncan, 2002). In terms of supporting these discourses and practices, health geographers have provided evidence, that physical, social and cultural environments affect population health in three ways: (i) by showing how poisonous contaminants – such as airborne and waterborne pollution – vary across space (Briggs et al., 1997); (ii) by showing how health-promoting resources vary across space and between social groups (Twigg and Cooper, 2010); and (iii) by showing how health-reducing social circumstances – such as deprivation, disorder and poor housing (Curtis and Jones, 1998; Dunn, 2000) – may 'amplify' unhealthy behaviours including smoking (Miles, 2006), alcohol consumption (Hay et al., 2009), drug use (Maas et al., 2007), poor and over eating (Cummins and Macintyre, 2006; Fraser and Edwards, 2010), and not exercising (Ellaway et al., 2007).

In terms of critiquing these public health discourses and practices, health geographers have carefully examined the roles and techniques of governments, key institutions and other officialdom in initiatives (Brown and Duncan, 2002). In particular, they have challenged the implicit assumption that health promotion is value neutral and devoid of politics, highlighting the role of soft paternalism and power (Brown and Burgess Watson, 2010), the active mobilization of concepts such 'risk' and 'empowerment' and ultimately the othering of certain bodies and groups in these processes (Keil and Ali, 2007; Ali and Keil, 2007; Brown and Burges Watson, 2010). This research is often based on exploring local circumstances (Wilton and DeVerteuil, 2006; Thompson et al., 2007) but increasingly extends out across interrelated scales to examine global public health discourses and practices (Brown and Bell, 2007, 2008).

The care of ill-health/production of health: services and institutions

Health care – i.e. the various sectors and institutions that attempt to deliver individuals from states of ill health to states of good health – has been a central concern of medical and health geography. Within this field of study, a longstanding predominantly quantitative research tradition has concerned itself with aerial differentiation and patterning, reporting how health care facilities, services and other resources are spatially distributed across local, regional, national and international scales, and the consequences of these distributions on service use and health outcomes (see Joseph and Phillips, 1984). Researchers have noted for example the 'friction of distance', and a 'distance decay' in service utilization (i.e. increasing distance equates to decreasing usage) (Stock, 1983). Notably, this research is based on an ethical assumption that it is a morally correct action to maximize accessibility to health services, and that part of this involves working towards an optimum spatial allocation of them (Meade and Earickson, 2000), indeed this being part of a broader sub-disciplinary concern for addressing 'underserviced populations'.

Whilst these interests and perspectives on accessibility and utilization have remained strong in the sub-discipline, since the early 1990s, moving beyond considerations of the shape and use of services, studies have paid much closer attention to features that form provision, such as administrative boundaries and local markets (Joseph and Chalmers, 1996; Moon and Brown, 2000; Cloutier-Fisher and Skinner, 2006) and health policy and regulation (Joseph and Kearns, 1996; Norris, 1997; Moon and North, 2000; Williams, 2006). These developments – often framed by a political economy theoretical perspective – have arguably led to this particular field of health geography being recognized as a body of work that contributes directly to mainstream health service debates on rationing, efficiency and equity in service planning and provision.

As Gesler and Kearns (2002) observe, the past twenty years have witnessed the infusion of neoliberal thinking, policies and structures into health care, and health geography has very much reacted to this, concerning itself with how this has played out in, and impacted upon, places. In terms of research, there has been a sustained interest in health geography on the broad transformations health care settings have gone through over time as a result of these forces: in other words, an interest in what powerful economic and political interests consider health care settings should be and represent, and what they have done to make this vision a reality. In much of this work, scholars have approached health care landscapes as 'texts' that they can read and decode to construct arguments on their meaning and significance (Kearns and Barnett, 1997). In particular, studies have argued how corporate principles in health care translate spatially into 'consumption landscapes' (Gesler and Kearns,

2002), whereby either health care institutions allow the market to colonize them or they, more thoroughly, embrace and manipulate their own market position (Kearns and Barnett, 1992; Moon and Brown, 1998). In terms of the market colonizing health care landscapes, an emerging move has been, through thoughtful physical design, to manipulate hospitals so that they feel, and are, warmer and more therapeutic to those who frequent them (Gesler, 2003; Gesler et al., 2004; Curtis et al., 2007; Evans et al., 2009). Beyond this health care spaces are increasingly being opened up directly to commercial enterprise to make them more exciting, engaging and even adventurous. Studies, for example, have considered conflict in the locating of fast food restaurants and shops in hospital atriums (Kearns and Barnett, 1999, 2000; Adams et al., 2010).

With regard to embracing and manipulating their market position, research has shown how hospitals themselves increasingly deploy commercial language and, through corporate branding and marketing, their own self-promotional strategies. Here, for example, the intent of management can be to de-emphasize negative identity with institutional medicine and, at the same time, promote achievements in positive ways to donors, politicians, private business, potential clients and the general public (Kearns and Barnett, 1999, 2000; Kearns et al., 2003; Moon et al., 2005, 2006; Joseph et al., 2009). Indeed, geographers have talked about how, using these self-promotional strategies, health care institutions have become locally, nationally and internationally 'famous' even when, as in the case of mental health asylums, they might have had controversial histories (Kearns et al., 2003; Moon et al., 2005, 2006; Joseph et al., 2009). On a related note, scholars have also articulated how communities develop attachments to, and responsibility for, their local hospitals and, through an identity politics and activism, oppose officialdom in support of them, particularly when facilities are threatened with cuts or closure (Moon and Brown, 2001; Brown, 2003; Barnett and Barnett, 2003).

The care of ill-health/production of health: workers and their work

An emerging strand of research in health geography is concerned with health care work and the health care workforce more broadly: the people whose job it is to deliver individuals from states of ill-health to states of better health, or at least assist them in living in ill-health (Andrews and Evans, 2008; Connell and Walton-Roberts, 2016). With regard to the informal sector, this research focuses on providers and users of care located in community settings such as homes (Dyck et al., 2005; Milligan, 2000, 2009; Bender et al., 2010) and streets (Conradson, 2003; Conradson and Moon, 2009; Wilton and DeVerteuil, 2006; Parr, 2008; Evans, 2011), including the power dynamics in play in place. It increasingly stresses the 'trans-scaled' and 'transnational' nature of this sector whereby carers care across diverse settings, relocate internationally to care or care at great distances (Wiles, 2003; Milligan and Wiles, 2010). Articulating experiences, this research sheds light on situations and circumstances in often less visible settings and roles, so that they are afforded the attention they deserve.

In the realm of professional care, one group of studies is focused on decision makers. It considers the geographical dimensions to, or consequences of, decision making across cohorts of workers at the macro scale (for example, family or hospital-based doctors or service managers). These decisions are often narrowly defined in nature, involving specific financial, planning or clinical concerns, and are often reactions to policy and/or broad system changes (Carr-Hill et al., 1994; Moon et al., 2002; Iredale et al., 2005). Also at the macro scale, broad population-based studies focus on the supply of labour, career decision making, and the consequences of these for local communities (Barnett, 1988, 1991; Cutchin, 1997; Baer, 2003; Laditka, 2004). In more recent years however, this macro-scale perspective has been complemented by a place-sensitive micro-scale perspective on the nature of work and workplaces. Topics considered here have included for example hospital strategic management (Hanlon, 2001), language in hospitals (Gesler, 1999), and the interpersonal and spatial dynamics that make and characterize specific specialities including general/family practice (Rapport et al., 2006, 2007), labour and delivery (Burges Watson et al., 2007), neonatal intensive care (Brown and Middleton, 2005), mobile dialysis (Lehoux et al., 2007) and mental health care (Curtis et al., 2007).

An important and substantive contribution to the geographical study of health care work has focused specifically on nursing (see Andrews, 2006, 2016). In this literature, the geographical qualities of the main patient and population groups nurses care for are considered (including where they live and how this is related to their use of nursing services; Hodgins and Wuest, 2007; Vandemark, 2007; Thomas, 2013). Elsewhere studies get to grips fundamentally with understanding how the job category 'nurse', and the numerous activities involved in the agency of nursing, relate dynamically to places (see Carolan et al., 2006; Solberg and Way, 2007; Stuart et al., 2008). A range of relationships have been articulated including how places possess meaning, symbolism and identity for nurses that are sometimes contested (Halford and Leonard, 2003; Cheek, 2004; Gilmour, 2006), and thus how places characterize particular professional sub-specialties of nursing (Burges Watson et al., 2007; Cheek, 2004). Meanwhile, further studies have addressed the nature of different places and their impacts upon nurse–patient interactions and relationships (Liaschenko, 1997; Malone, 2003; Peter, 2002; Liaschenko et al., 2011), and upon inter-professional interactions and relationships (Barnes and Rudge, 2005; West and Barron, 2005; Halford and Leonard, 2003).

Feelings on journeys to health: therapeutic and healing experiences

The notion of 'therapeutic' has been subject to multiple and often very broad definitions across academic disciplines and clinical specialties. However, for a great deal of research across the social sciences therapeutic is used to describe a particular feeling, related variously to experiences such as healing, relaxation, restoration, contentment and being at peace. It is this understanding of therapeutic that health geographers have spoken to when developing the idea of therapeutic landscapes. This concept was first introduced to geographers in a position paper by Wil Gesler over twenty-five years ago (Gesler, 1992). Taking a broad theoretical sweep, Gesler drew on and blended humanistic geography (the ideas of sense of place and symbolic places) and structuralism (the ideas of territoriality and hegemony) in his arguments for the development of research that would pay attention to the psychological, social, cultural and economic processes in how the healing process works out in place, and in how certain places become known for their healing qualities. Although there are many ways of categorizing the empirical interests of therapeutic landscape research, broadly speaking two person-focused themes have emerged alongside three place-focused ones. With regard to the former, research has considered different life contexts and changes through which therapeutic landscapes are pursued or encountered. For example, studies have considered the experiences of migrants and refugees and specifically the importance of place-making practices in the creation of therapeutic landscapes (e.g. Gastaldo et al., 2004; Sampson and Gifford, 2010). Research in this vein has also considered the formation of therapeutic landscapes that act to support people through emotional experiences associated with illness and their corresponding treatment journeys (e.g. Donovan and Williams, 2007; English et al., 2008). On another level, research has considered different belief systems that map onto place and give rise to therapeutic landscapes. Studies have considered, for example, the cultural health norms of developing-world peoples and populations (e.g. Sperling and Decker, 2007), of First Nations/ indigenous people and populations (e.g. Wilson, 2003; Wendt and Gone, 2012), and the theories and philosophies underpinning holistic medicine and lifestyles (e.g. Andrews, 2004; Hoyez, 2007a, 2007b; Lea, 2008).

Of the three place-focused themes, the first explores what might be broadly termed 'natural environments'. Studies in this area consider the convergence between settings that support some degree of engagement with nature and the promotion of health and healing (e.g. parks and gardens, woodland and wilderness spaces; Palka, 1999; Milligan et al., 2004; Milligan and Bingley, 2007), there being a close link between this research and the more general focus in human geography upon the therapeutic qualities of green spaces (e.g. Maas et al., 2006; Lee and Maheswaran, 2011; Ward-Thompson and Aspinall, 2011). The second place-focused theme, 'designed and built spaces', is concerned with the creation of therapeutic landscapes through medical and political ideas, economic conditions and private sector ideology and enterprise. Particular attention has been paid here to the promotion of healing through improvements in design, for example, through the use of colour, light and art in hospitals (e.g. Gesler et al., 2004; Gesler and Curtis, 2007; Curtis et al., 2007) and primary health care spaces (Crooks and Evans, 2007; Evans et al., 2009). Outside of health care, the importance of design to therapeutic landscapes has also been explored through attention to a range of other settings including children's health camps (Kearns and Collins, 2000), former mental asylums (Moon et al., 2006), homes (Williams, 2002; Donovan and Williams, 2007), spiritual retreats (Gesler, 1996; Conradson, 2007; Williams, 2010), through to work on neighbourhoods and towns as therapeutic landscapes (Andrews and Kearns, 2005; Wilton and DeVerteuil, 2006; Foley et al., 2011). A third and final place-focused theme extends the idea of therapeutic landscapes even further; that is, by focusing on 'imagined places'. Here, the emphasis is on places that are not necessarily materially present but are visualized often for the purpose of escape, relaxation or to promote psychological resilience. Studies in this area have considered, for example, places evoked in fictional writing (Gesler, 2000; Baer and Gesler, 2004; Tonnellier and Curtis, 2005; Williams, 2007) in soundscapes in popular music (Andrews et al., 2011; Evans, 2014) as well as in personal memories (e.g. Gastaldo et al., 2004).

Most recently, a new era of therapeutic landscape research has made substantial critical and theoretical progress. At one level, studies have addressed the previous lack of attention to known processes and sectionalities in society such as gender relations (Love et al., 2012; Little, 2013), and global connections and relations (Hoyez, 2007a,2007b). At another level, studies have considered the anti-therapeutic effects of certain places and/or the possibility that unplanned, dangerous and unpleasant experiences can actually be part of the overall experience (Milligan and Bingley, 2007; DeVerteuil and Andrews, 2007; Andrews and Holmes, 2007; Laws, 2009; Willis, 2009; Smith et al., 2010; Wood et al., 2013). This era of research has also seen some precise attention to how therapeutic landscapes might arise and work. Studies have considered for example media mechanisms and roles (MacKian, 2008), psychoanalytic explanations and psychological processes (Evans et al., 2009; Rose, 2012; Conradson, 2005b), and the importance of individual and shared movement (Doughty, 2013; Gatrell, 2013).

Health as something, and something meaningful: engagements with wellbeing

Playing their parts in a broader 'wellbeing turn' in human geography, health geographers have helped demonstrate multiple ways in which wellbeing is defined and utilized in contemporary society. For example, at one level how wellbeing surfaces in the language and statements of policy makers as something that can be structurally provided for, achieved and maintained. The associated materialist understanding being that improving a population's financial prosperity, and the range and quality of goods and services it can access, increases its wellbeing (MacKian, 2009; Smith and Pain, 2010; Atkinson et al., 2012; Schwanen and Atkinson, 2015). Health geographers have also acknowledged the very different ideas that exist on how wellbeing might be accomplished, ranging from the political left's mobilization of the idea of 'social wellbeing' to bolster arguments for the development of welfare systems and structures, to the political right's mobilization of the idea of 'economic wellbeing' to support the neoliberal agenda towards marketization and the distribution of resources through private enterprise (Fleuret and Atkinson, 2007; MacKian, 2009; Atkinson et al., 2012; Schwanen and Atkinson, 2015). Moreover health geographers have acknowledged a broad societal 'will to health' – part of the self-help movement's belief that wellbeing results from living a good life through attention to spiritual and bodily needs (MacKian, 2009; Kearns and Andrews, 2010) – and have acknowledged how wellbeing arises somewhat organically and free of identity agendas in everyday contexts, contacts and activities (Schwanen and Wang, 2014; Andrews et al., 2014).

Recent critical engagements with wellbeing have intensified in health geography with a small number of scholars beginning to consider wellbeing far more directly, critically and theoretically by asking two fundamental questions, the first of these being: what is wellbeing and how does it occur? In terms of answers, a consensus seems to be emerging that wellbeing is attained in some way by fulfilling personal needs – vital, spiritual, emotional and materialistic (see Fleuret and Atkinson, 2007; MacKian, 2009; Aslam and Corrado, 2011) – whilst acknowledging human variability either in terms of desires, values and preferences (realized in choices), or in terms of capacities and restrictions (realized as limitations) (Fleuret and Atkinson, 2007). Meanwhile consideration is given in research to specific approaches and routes to wellbeing such as through activism (Muirhead, 2012; Wheeler, 2012), nationhood and international development (Schaaf, 2012), and how political interests construct wellbeing and actively use it in governance (Atkinson and Joyce, 2011; Scott, 2015; Foo et al., 2015). Other studies have considered fundamental theoretical and compositional issues such as subjective vs objective wellbeing and their inter-relationships (Schwanen and Wang, 2014; Fleuret and Prugneau, 2015), stable feelings of wellbeing vs momentary feelings of wellbeing and their inter-relationships (Schwanen and Wang, 2014), relational vs situational facets of wellbeing and individual vs collective facets of wellbeing (Atkinson, 2013). The second critical question is: where do different forms of wellbeing arise and how does space and/or the nature of place play a role in such situations? (Fleuret and Atkinson, 2007; Atkinson et al., 2012). In terms of answers, Fleuret and Atkinson (2007), for example, highlight spaces of human capacity (that assist and amplify wellbeing), integrative spaces of networks (that spread wellbeing), spaces of security (that provide refuge and support) and therapeutic spaces (that facilitate healing). Moreover, a range of studies also consider various qualities of place that are thought to produce or change wellbeing (Beck, 2012; Gilroy, 2012; Riva and Curtis, 2012).

The movement of research in health geography into studying therapeutic and wellbeing experiences constitutes an important step towards a more substantial sub-disciplinary engagement with health. Indeed, for the first time it edges the understanding of health beyond being a state or level where disease is absent, to where it is recognized as something in itself, including feeling and becoming more physically and mentally healthy.

Towards a more-than-representational idea of health

As the above summary sections illustrate, over the years health geography has viewed health through positivistic, political economy, social constructionist and post-structuralist theoretical lenses (which illuminate how it is distributed, resourced, constructed, and negotiated). Moreover, the sub-discipline has recognized health as a state that can be reduced (through disease), maintained (through public health), restored/obtained/bought and sold (through care), or something more complex and experiential in itself that might be felt and cherished (through healing and wellbeing).

How we might move beyond all these representations – measurements and meanings – and towards what, is proposed by Cameron Duff in his book *Assemblages of Health* (Duff, 2014a). Drawing on the writings of Deleuze, Duff's scholarship is focused on the basis and mechanisms for health. Duff grounds his arguments for reformed thinking on health in some fundamental observations on the changing nature of society and human life. First, that under technological developments (such as genetics), claims to previously firm distinctions and counterpoints – such as cells vs society, nature vs culture, natural vs artificial, biological vs machine – are becoming increasingly tenuous. Duff's point here being that if human life is itself being re-conceived, then research approaches also need to be re-conceived, offering an account of health that pulls down and redraws boundaries between the human and non-human, and is more aware of the multiple overlaps and relations between them. Second, Duff also points out that although what health is has always been a complex question, in the twenty-first century it is ever more so because numerous moral, biological, genetic, psychological, cultural, political and economic phenomena increasingly come into play that give it different connotations. Moreover, he argues that, although academic, policy and practice efforts are increasingly being made to diversify thinking on health, these have in common their subjectivity and their sidelining of fundamental discussions around health by replacing them with numerous other concepts with their own definitional and implementation challenges (e.g. 'functionality', 'fitness', 'resiliency', 'thriving', 'wellbeing', 'happiness', 'quality of life' and so on). Hence Duff argues that, in the face of such diversity, academics might be better off getting back to basics by, rather than thinking about what health is, thinking about how health arises.

In addition to developing the aforementioned general critiques, Duff has some quite scathing criticisms of mainstream population health research. Whilst he acknowledges that its literature on social determinants of health is vast, he suggests that one might challenge the whole notion of social determinants as being objective, stable, definable realities. Moreover, he suggests that one might challenge the causal links established between 'determinants' and health inequalities (and the regarding of these as discrete). For Duff, at issue is the reliance on a questionable logic and inching agenda whereby social structures and processes – as suggested previously, often identified as compositional, contextual and collective characteristics of places – have been expanded to include almost all human circumstances and behaviours (e.g. employment, crime, wealth, cultural norms, built environments, and even social capital) and quantified as part of an ongoing mission to reduce more and more of them to statistical levels. Indeed, as Duff reminds us, and as Deleuze before him argues, nothing is determined in life; nothing is absolutely closed and linear.

In terms of a way forward, Duff argues that NRT and related thinking needs to be part of a new ethology or 'minor science' of health that rejects the ontological distinction between animal and human bodies as well as between artificial and natural objects:

> Ethology is first of all the study of the relations of speed and slowness, of the capacities for affecting and being affected that characterize each thing. For each thing these relations and capacities have an amplitude, thresholds (maximum and minimum) and variations or transformations that are peculiar to them. And they select, in the world or in Nature, that which corresponds to the thing; that is, they select what affects or is affected by the thing, what moves or is moved by it … an animal, a thing, is never separable from its relations with the world … The speed or slowness of metabolisms, perceptions, actions and reactions link together to constitute a particular individual in the world.
>
> (Deleuze, 1992:627–628).

Indeed, as Andrews (2017b) and Duff (2014a) both outline, as an approach, this ethology might be able to establish a more robust and substantive idea of how health arises that:

(i)   Does not reproduce a divide between the subject/social (typically deconstructed via social sciences) and nature/biology (typically deconstructed by health sciences); a divide evident in social epidemiology which can only link them.

(ii)   Thus helps scholars escape the urge to either focus on individuals and populations (their experiences, practices, beliefs etc.) or on mechanisms (such as pathogens, discourses, power relationships).

(iii) Needs no specific presupposed criteria to determine the boundaries of health-related events, but instead follows the things/actors that reveal in the processual aspects of humans becoming more or less healthy.

(iv)   Does not see health as simply the natural state of a normal human body, but redefines health as a state or experience of a particular modulation of life produced within particular assemblages (thus, by implication, pays attention to the varied origins and character of health).

(v) Seeks a post-human account of health which is sensitive to the workings of these health-mediating assemblages; the human and non-human entities and the many mechanisms (affects, relations and events) involved in becomings in particular places.

(vi) Rather than attributing health to structural impositions or transcendental phenomena (such as rules, laws, god, reason or nature), instead recognizes health arriving within a self-organizing process that constitutes the manifestation of the material world. Thus recognizes health arriving on a 'plane of immanence' within a multiplicity of human bodies, objects and forces all operating on the same level of existence.

(vii) Acknowledges that health is not stable or static, that health states and experiences are only momentary actualizations, and that human bodies are always in the process of becoming more or less healthy.

(viii) Has an experimental mode, typical of a minor science, which looks for flux and problems in knowledge (including the conventions of state science) and embraces non-scientific activity as part of methods and solutions.

Notably, as the above points suggest, such an ethology has to recognize the body differently, as 'active and open'. Indeed, whilst traditional positivistic medical geographies see bodies as units (such as infected or not infected, moved or stationary), humanistic health geographies see emotional and physically feeling bodies (experiencing, negotiating, coping and articulating), and post-structuralist health geographies see bodies as ordered (categorized and regulated by medical institutions), as Greenhough (2011a) suggests, an NRT informed ethology instead draws on ideas such as Deleuze and Guattari's (1988, 2003) 'body without organs' to see bodies as collections of matter and energy with multiple potentials (including to resist processes imposed by the medical gaze). Health then being a 'more-than-human' accomplishment involving the body constantly in the process of becoming (more of the same or something else) through its encounters with other bodies and objects (see also Fox, 2011). As Gagnon and Holmes (2016) suggest, Deleuze and Guattari (1988) challenge our very understanding of the body by insisting that the body has no meaning in itself; it has no 'essence' and cannot be defined as a single physical unit, existing instead in the form of an open surface able to connect with other bodies and a multitude of heterogeneous elements. The key question for health research hence becomes: how might the body be exposed to more positive forces of becoming (Duff, 2014a)? Moreover, in terms of the translation and impact of this research, the priority becomes to adopt an ethics that aims to maximise positive action; to encourage these positive forces and the body's movement within them (Duff 2014a) (see also Chapter 8).

Assemblage theory

> An assemblage [is] every constellation of singularities and traits deducted from the flow – selected, organized, stratified – in such a way as to converge ... artificially and naturally; an assemblage, in this sense, is a veritable invention.
>
> (Deleuze and Guattari, 1988:406)

The term 'assemblage' is currently deployed quite liberally across the social sciences, mirroring an aim amongst scholars to describe what composes events and phenomena in places in a way that is anti-structural, yet still speaks to their content and some order (Marcus and Saka, 2006). There are, however, as Greenhough (2011a) suggests, quite focused ideas in NRT on assemblages (which are also central to Duff's ideas of an ethology of health, and the making of health). NRT recognizes the strong theoretical basis of assemblage theory in the work of Deleuze in particular, here it being an ontological framework, borrowing from dynamic systems theory and other traditions, for explaining social composition, complexity, functioning, fluidity and drive (see Deleuze and Guattari, 1988; DeLanda, 2006; Dewsbury, 2011; Duff, 2014a, 2014b). Deleuzian assemblage theory argues that assemblages arise in every place where life arises and are composed of all bodies and objects (elements) in place and/or in the process of moving in and out of place, their relations, events and affects constituting the basis and emergence for all life. Indeed, in two operations which are distinct yet occur almost simultaneously, assemblages are 'created' through a process involving the grouping of bodies and objects, and 'actualized' through the release of the potential of these bodies and objects (Buchanan, 2000; Duff, 2014a).

Moreover, as Little (2016) argues, whether used within and beyond NRT, the basic principles of assemblage theory include that: (i) social entities are systems composed of ever lesser/smaller elements; (ii) elements of a social entity are heterogeneous with their own character and dynamics and material and expressive roles; (iii) any assemblage draws elements from different temporal and spatial scales with their own forces and affects which help it continually form or deform, and it express similar or new forces and affects (as Duff (2014a) suggests, assemblages ensnare all life in a network of relations, but relations ensure the openness of the assemblage to new relations that might transform it); (iv) interactions amongst elements may be indeterminate because of complexity in terms of volume and causal mechanisms; (v) the behaviour of the whole assemblage is impossible to calculate even given extensive knowledge of the elements and their dynamics. Further, providing more specific detail on how this all works out, DeLanda (2006) argues that assemblages are a product of historical processes which 'rehearse' the capacities of elements; elements being involved in processes of territorialization/coding (which orders and stabilizes the assemblage and potentially its identity) or deterritorization/decoding (which does the opposite). Moreover, assemblages are territorialized both in terms of their material content (including, for example, human and non-human bodies, actions, and reactions), and their non-material expression (including, for example, incorporeal enunciations, acts, and statements).

In sum, assemblage theory describes a world borne in bodies and objects which are never detached or discrete but are contingent, co-dependent and relational. A world where nothing is stable, fully formed or complete, only fluid and emerging. A world where no permanent wall of separation and distinction exists between one kind of being and another (e.g. human and micro-organism). A world where, because of the activity within assemblages, phenomena can be, and feel, more than the sum of their parts (most of these understandings contrasting, for example, with a world based on language, subjectivity and meaning). NRT does not often, if at all, strictly rehearse the aforementioned processes, yet they form a basis for its understandings of, and inquiries into, the happenings in life, scholars asking fundamental questions of assemblages such as: What is present? What is arriving? What is leaving? What is passive? What is active? What is interacting, how and with what? Applying this understanding to the health creating or detracting process might inform any number of substantive empirical fields of inquiry (disease, health care, public health etc.). Indeed, each involves particular assemblages that either create or reduce health; assemblages that often collide, conflict and contradict in terms of their workings. Although most health geography informed by NRT is consistent with assemblage thinking and theory, the following two sections outline empirical areas where it is at the forefront of studies.

Assemblages in more-than-representational health geographies: wellbeing emerging in community contexts

An NRT-informed understanding of wellbeing has developed in health geography that moves thinking away from the aforementioned cognitivist, economic and humanist assumptions and the proxies often deployed (such as wealth, consumption), the 'being' in wellbeing instead understood as exposures to, and interactions with, components of assemblages which act as a causal network (Atkinson, 2013; Andrews et al., 2014). Such an understanding accounts for underlying processes and the ways in which the physical, biological, neurological and cultural combine simultaneously in the processes of wellbeing emerging. Moreover, it conceives wellbeing as something unstable and amenable to immediate change, something both individual and collective, something both consciously and less-than-fully consciously known, thus as something both subjective and objective (Atkinson, 2013). In research terms this begs the production of highly relational and situated accounts: wellbeing as performed and often shared, rather than as an individually and gradually acquired attribute (Atkinson and Rubidge, 2013; Atkinson, 2013; Andrews et al., 2014).

In terms of emerging empirical research, three interrelated themes can be traced in the current literature, the first being 'healing and recovery'. Duff (2011), for example, represents a relational model of enabling places for mental healing and recovery, describing assemblages here emerging from a web of connections/associations and processes involving social resources (e.g. social capital, connections, trust), affective resources (the feeling of people engaging with a place and its activity) and material resources (e.g. parts of education, entertainment, transport). In his later paper Duff (2012) tests his 2011 model with a qualitative study. With regard to social resources, a respondent, for example, talks about local friendship networks:

> It's a very important thing in life to be connected with people who understand what you are going through. Being connected to friends is really important and now my friends know um, just where things are with my life. They are all really sensitive about it and that makes a huge difference, just like day to day.
>
> (2012:1391)

With regard to affective resources, a respondent, for example, talks about walking at night:

> Going for a walk at three o'clock in the morning, I feel quite safe on the street, cause its dark and there's no one around, everyone else is asleep. There's no energy floating around the air, you know people's manic energy, everybody's resting. So I find the streets around here quite calm and peaceful at night, even though it goes against what should be because you're not supposed to feel safe at night. I don't know but for me the darkness is safe. The world is at rest and it just makes me calmer.
>
> (2012:1391)

With regard to material resources, a respondent, for example, talks about his shed:

> My shed holds my treasures. It's masculine too, like I build my life around the shed, it's part of my life. Like I work in my shed, I work on my recovery.
>
> (2012:1392)

Meanwhile, working to similar principles, Foley (2011) considers a specific focal point as the basis for an assemblage in healing and recovery; that of the holy well. Indeed, in his study Foley attempts to reconceptualize the idea of therapeutic landscapes in assemblage terms. The holy well assemblage, he argues, is constituted of structures that are material (felt by the body), metaphorical (envisioned/believed) and inhabited (lived/performed). Similarly, in his later work on the Roman-Irish bath Foley (2014) extends assemblage thinking in therapeutic landscapes further by examining this particular assemblage's networking and diffusion across the Victorian world and beyond in time and space, as a socio-material phenomenon and practice. Specifically, he suggests:

> [T]he assemblage of the Roman Irish Bath can be seen as a relational entity: a container (of sweating cures, therapeutic design and medico-moral practice) but also a distributor (through economic expansion and reproduction).
>
> (2014:17)

The second theme in this literature is the facilitation and constraint of well-being through assemblages (see Malins, 2004; Duff, 2012). Focused on disability, for example, Stephens et al. (2015) suggest that the traditional social model of disability rightly aligns physical absence with social exclusion yet it risks wrongly conflating physical presence and social inclusion. Perhaps more important, they suggest, is what the body can do – what affects it can pick up and transmit – once in place. Thus, Stephens et al. advocate for a research understanding that escapes the aforementioned associations and is focused more generally on bodies and social contexts in a way that retrieves and re-centres the (dis)abled body in its analysis, emphasizing corporeal capacities and experiences in and amongst built environments, and the moving bodies and objects involved. In particular, the authors suggest that scholars, and ultimately policy makers, might distinguish between 'oppressive' and 'emancipatory' assemblages based on the ways they constrain or enhance bodily capacities (see also Ruddick, 2012). An example from their empirical study shows how a non-human/object addition to a caring assemblage does not have the desired affect for a disabled child:

MOM: They ordered a table specifically for her … it looks like the others but it has adjustable legs.

CHILD: Now it doesn't stop squeaking, when I move it doesn't stop squeaking. I don't want to move because every time I move it makes sound.

(Stephens, 2015:207)

A final example of research in the second theme is Duff (2014b), who flips his aforementioned resource model in order to articulate the social, affective and material resources involved in assemblages of urban drug use – hence to understand the basis of an activity that ultimately detracts from wellbeing (see also Duff, 2016b).

The third theme in this literature is movement in assemblages for wellbeing, these accounts finding wellbeing in the actions and vectors which constitute different space-times (i.e. space-times made by walking, dancing, playing, caring etc.) rather than wellbeing in discrete subjects (see Atkinson and Rubidge, 2013; Atkinson, 2013). Here the role of technologies in assemblages has been a particular focus of attention. Middleton (2010), for example, conceptualizes walking in everyday life as a socio-technical assemblage (technologies including pavements, walkways, lights, vehicles as hazards etc.). Thinking of walking in this way, she suggests, allows attention to be placed on embodied, material and technological relations and their consequence for daily urban movements made on foot. Similarly, but with more of a micro-scale perspective, Barratt (2011) conceptualizes recreational climbing – an activity undertaken for a mixture of fitness/health, wellbeing, sensory and competitive reasons – as a foot-shoe-rock-based assemblage within which ever present technologies enhance the capacities of bodies to perform.

Assemblages in more-than-representational health geographies: behaviours, diseases and their institutional entanglements

Research in health geography considers the assemblages that constitute and surround ill-health, including its various system engagements. One strand of scholarship is focused specifically on infectious disease. As a point of departure, as suggested earlier, it is noted here that most conventional epidemiological research does not actually focus on pathogens, viruses, bacteria and vectors themselves, but is more to do with their downstream impacts on human populations (Mayer, 2010; Greenhough, 2011a). It is argued, however, that through NRT-informed inquiries, research might instead focus on these things and their agency and roles within wider assemblages and networks that constitute disease events. Indeed, it is shown in studies how pathogens, viruses and bacteria are coinvolved with other living and non-living participants (bodies, ecosystems, pollutants, infective agents, health care institutions etc.) which together provide an assemblage of more or less 'pathogenicity' determining the spread or retreat of disease (Keil and Ali, 2006, 2007; Greenhough, 2012; Hinchliffe and Ward, 2014; Hinchliffe et al., 2016). Taking these ideas off in an interesting direction, Greenhough (2011a), for example, uses assemblage theory to consider a proposal to turn Iceland into a laboratory of gene research, in particular asking: how do certain life science and other assemblages gain friction (notoriety, focused attention) when we know all assemblages to be temporary, unfinished, and ever reconstituting? In summarizing her view of how these assemblages work and the value of assemblage theory more generally, Greenhough quotes Foucault:

> Being a collective phenomenon ... [an epidemic] requires a multiple gaze; a unique process, it must be described in terms of its special, accidental, unexpected qualities.
>
> (Foucault, 1997:25)

Another strand of research is concerned with how institutional power is exerted internally in particular settings. Sullivan (2012), for example, considers how inequality is enacted through global and state governance within a Tanzanian hospital. Specifically, the author notes that the unequal material layered configurations that characterize hospital spaces mean that some individuals can attain meaningful identities as patients and health professionals whereas others cannot. The author argues:

> If the hospital is a place where the state and the global are produced through assemblage, it should not be assumed that the entire hospital as an institution is necessarily linked up in assemblage, or that those hospital spaces that are made global are necessarily stable. Conversely, the [his] ethnography demonstrates that assemblage is punctuated, contingent, and emergent within hospital spaces; certain spaces are made global through state/donor governance regimes, while others – by virtue of being beyond state/donor interest – remain decidedly local. Thus, the kinds of ordering processes performed within global and local hospital spaces are often incongruous, with important implications for how people move through the hospital.
>
> (2012:59)

In another study, focused on the semi-institutional setting of school, Rich (2010) considers surveillances in the production of obesity as a category. Applying Haggerty and Ericson's concept of the 'surveillant assemblage', the research examines how surveillant practices in schools are part of an assemblage constituted by a range of agencies/officialdom, and socio-technological developments.

Elsewhere research shows how institutions and their initiatives can work in positive ways; which is perhaps not surprising given that this is almost always their goal. Atkinson and Scott (2015) study a primary school based dance and movement intervention (typically categorized amongst more adventurous thinkers on school curriculum as 'non-sedentary' approaches to learning and wellbeing). Here, the authors found that the normal classroom assemblage was altered during the intervention, the daily status quo being disrupted through adding musical sound, removing chairs and adding props. This, they argue, enabled more movement, provided opportunities for more expressive movement, escaped normal school-day habitual practices, and created an open space of affective possibility and action.

A final strand of research is concerned with how institutional power is extended into community space. Focused on HIV treatment programs, Gagnon and Holmes (2016) argue, for example, that assemblage theory offers a new way of theorizing the side effects of antiretroviral drugs, and ultimately the relationship between the body and these drugs (as technologies). Using examples, the authors examine the multiple ways in which the body connects, not only to drugs, but also to other objects and systems. Elsewhere, developing a framework for a critical geographical approach to global health, Brown et al. (2012) describes the three allied analytical approaches that constitute it: governmentality, risk and assemblage. With regard to the latter, the authors argue that assemblage helps account for the multiple actors and forces that combine through scales in any one place/ context (without reducing analysis to a single logic or cause), including those driven by capitalism, regulatory structures and socio-cultural norms and practices. Although published earlier, demonstrating Brown et al's ideas in a specific context, McCann (2011) focuses on the flows of policy and knowledge that occur between cities and around the globe that create the broad public health approach of 'harm reduction' (in the governance of illegal drug use), and how in particular places these are actively (re)assembled, inserted and negotiated. The author argues that using the idea of assemblage brings to their analysis a number of often neglected variables including urban policy making, urban politics, and global trends and movements.

In sum, the use of assemblage theory is expanding in health geography, it laying and important foundation for more-than-representational scholarship. More could be done in future however in terms of thinking about how assemblage theory might inform particular fields and debates in the sub-discipline.

3 Key conceptualizations in more-than-representational health geographies

This chapter describes some core theoretical concepts and ideas in NRT that help articulate how the immediacy of life and the human experience comes to be: relational materialisms, onflow, affect, vitality, virtuality and multiplicity, and hope (for other typologies see Thrift, 2000, 2008; Cadman, 2009; Vaninni, 2009; 2015a; Andrews, 2014a). The chapter elaborates how each has been bound to explanations of health in more-than-representational health geographies (as defined in the Preface and Chapter 1).

Relational materialities

As suggested in the last chapter, NRT asks: 'what can bodies do?' Equally, though, it also asks: 'with what objects?' and 'how do these objects act?' Indeed, a fundamental concern of NRT is to embrace relational materialism in thinking about how life is constituted and reproduces. This – possessing obvious commonalities and compatibilities with assemblage theory – involves a number of core understandings, that:

(i)   Being dynamic and possessing potential and agency, material objects are more than 'props' for humans (thus life progresses through the co-evolution of human bodies and non-human objects).

(ii)   Human bodies and non-human objects, and their qualities and capacities, do not exist alone, but only through their relations with, and in relation to, numerous other human bodies and non-human objects and their own particular qualities and capacities (thus life is physically and complexly networked, which is all critical to its making).

(iii) Life in any one place (and the bodies and objects that create it) is complexly networked with life in other places (and their bodies and objects) potentially across vast geographical distances. Thus, thinking of life in this way disrupts traditional notions of scale; life being trans-scaled.

(iv)   Due to never ending influxes of material and other influences into places, and constant interactions between the bodies and objects already in place, relations are never closed or complete. Hence places are forever temporary accomplishments, constantly becoming (constantly being negotiated, contested, reopened, reworked – changing and developing).

(v)   Due to these material relations operating everywhere, society itself becomes a set of networks in which expressive practice acts to produce either coherence or change.

(vi)   The fundamental form of material objects (e.g. sizes, shapes, textures, colours etc.) and their movement (e.g. speed, direction, momentum) are important to relations because they affect events and their outcomes.

(vii) The relational performances of material happenings (bodies and objects assembling and interacting) are important, it thus being necessary to think about the specificity and performative efficacy of different relations and different relational configurations (for wider discussion of these understandings see Kraftl and Horton, 2007; Darling, 2009; Jones, 2009; Anderson and Harrison, 2010).

Notably, quite different ideas on space arise from these seven ideas. Rather than conceive of space as an objective plane where distance between points should be measured, or as the basis for scale, NRT approaches the world through a 'flat ontology' (Marston et al., 2005). In other words, space does not exist apart from the human and non-human entities that make it. As Jones (2009:491) describes it, 'objects *are* space, space *is* objects, and moreover objects can be understood *only* in relation to other objects' (because objects move, this giving rise to the idea of 'space as a verb' common in NRT). Such an approach also brings to the fore new understandings of space-time. Indeed, rather than space-time being conceptualized as a backdrop *to* relational networks, configurations *of* space-time are understood to be produced *by* relational networks. Thus, it becomes important to think about the spacings and timings of things (Thrift, 2008).

In sum then, in contrast to traditional approaches in the social sciences, relational materialism constitutes a post-humanistic theoretical perspective that is skeptical of others that are person-centred, biographical, and that tend to isolate, narrow down and bracket phenomenon (see Thrift, 2008; Vannini, 2009). Being 'pre-individual', it is concerned with processes that are not primarily subject/person-based; the word 'I' not being common to the vocabulary used in empirical applications (Anderson and Harrison, 2010). As Duff (2014a) explains, relational materialism acknowledges the enduring complexity and flux of the world: that researchers cannot fully explain all of what is going on, yet can grasp the nature of the multiple contingencies, distances, associations and changes in the creative process of life. These being relations that exist 'in life' rather than resulting from, or being activated 'by life' (Duff, 2014a). So, as Anderson and Harrison (2010) suggest, in NRT it is important to study associations and mutuality – co-invention, co-evolution and co-fabrication – between things. Thinking again of word usage, 'and' and 'with' are far more important to scholars of NRT than 'is'.

Relational material health geographies

Some initial thinking about materiality and relationality in health geography has emerged in the consideration of how health-related scientific knowledge is constituted and produced. Hall's early work on human genetics is notable in this regard; Hall (2003), for example, considering gene mapping and the way, as both body and object, genetic material is transformed into a 'knowledge-able entity'. Shortly later Hall (2004), deploying actor-network theory (ANT) to think about the material, relational and ongoing nature of knowledge production in genetics and heart disease (the co-equal role of human and non-human actors, and their spatial networks). Notably, these early ideas on scientific enterprises are continued in Greenhough and Roe's (2011) recent work on animal laboratories which emphasizes the relationalities and co-becoming between lab workers and rats – i.e. human bodies and non-human bodies. Specifically this research articulates the way implicit and mutually reinforcing actions develop in these technological environments, the workers 'becoming' better technicians and the rats 'becoming' better at domesticated living.

Perhaps the strongest testing ground for material and relational thinking in health geography has, however, been the empirical field of complementary and alternative medicine (CAM), not only because of the theoretical orientation and preferences of the particular scholars who study it, but also because of the wide array of unusual materials and practices it involves, oftentimes with diverse historical and geographical origins. Certain research focuses on the relational roles of materials in CAM assemblages, Doel and Segrott (2004) for example creating a typology which emphasizes the expressive agency of 'signature materials', 'supplementary materials' and 'marginalia' that make each therapeutic encounter, in contrast to much conventional medicine, non-generic and tailored, often leading to unexpected and unique events and experiences. With regard to signature materials, Doel and Segrott's own empirical inquiries highlight the relationship between essential oils and human bodies in aromatherapy, they arguing that:

> Essential oils are highly prized for their therapeutic properties. These oils are central to the therapeutic encounter because they transform bodies, and enable particular kinds of embodied experience to take place... through the tactility of massage and the aromas of essential oils, the experience of aromatherapy may become a sensuous, emotional, and even uplifting event in its own right.
>
> (2004:730)

Other research fits neatly into Doel and Segrott's typology. With regard to supplementary materials, Andrews (2004) highlights the importance CAM therapists place on manipulating their own practice space to help create a desired ambience and atmosphere. In this study, a massage therapist/

reflexologist comments on her interior design and decoration that included visual and audio stimulation:

> As you can see, my clinic is decorated in a way which helps clients drift away. I try and make it a relaxing place with soothing images and sounds. I don't put a lot of effort into it, I just buy things that I think will add to the atmosphere.
>
> (2004:313)

With regard to marginalia, Andrews et al. (2013) describes how therapists actively exclude certain objects and related practices from the therapeutic encounter, even if they might be critical to their own business more generally and their clients' lives. In their study, a therapist describes their views and personal policies on this topic:

> We aim for a basic life level, like no computer in the room. For them to not hear phones ringing and faxes going, and see computer screens. It lets them exhale and get more in tune with what's really going on with themselves. They can hear themselves again. I don't want them to have demands so they have to turn their cell phones off. Its nurturing, and I think that's really a big difference. When they walk into a calming, quiet, warm, sort of down to earth place, they can come into themselves, reflect a little bit about themselves.
>
> (2013:105)

At the same time, research on CAM has also focused on the roles of human bodies and minds in the CAM assemblage, demonstrating the core value of their movements and judgements in the relationalities involved, and thus to the therapeutic experiences that emerge. Using ANT, Andrews et al. (2013), for example, show how therapists contribute to CAM actor-networks by being mediators who stabilize associations among materials/objects through processes of translation to other human actors (this having the effect of extending fields of identification with and within the CAM network as more actors align their interests). Moreover, Andrews et al. (2013) describe how three moments are important to therapists' translation processes: 'procurement', 'arrangement' and 'guidance'. With regard to procurement, a therapist, for example, says:

> I picked up some excellent things whilst travelling in India, mainly Hindu and pieces and artisan pieces that I used to decorate my practice space. I bought some CDs of local ambient music that mix traditional instruments and modern electronic.
>
> (2013:103)

With regard to arrangement, the researcher's own observational field note, for example, reflects:

> This larger clinic is different in that therapists don't always have the same or their own practice space. As a result the treatment rooms are not personalised and many are slightly sterile, not what you'd expect in CAM. What stuck me however was that this did not seem to be an obstacle for some of the therapists who literally brought their environment with them. One for example, had a bag complete with photographs of tropical places, ornaments, books, small electrical fan and a sea breeze air refresher. A kind of 'clinic in a bag' that she cheeringly but subtly arranged each time, not making the arranging itself the focus of attention.
>
> (2013:104)

Meanwhile, with regard to guidance, a therapist, for example, states:

> Its about showing people that medicine [bio-medicine] is not the only solution and often not a solution at all, and that people in other parts of the world have long been practicing in ways that can help health and vitality. They learn much more on top of that because therapies don't stop at treatments. They are about other ways of living your life and having other priorities. It's not all about shopping and working.
>
> (2013:104)

Beyond CAM, two other empirical fields within which an explicitly relational materialist analysis has been forged, are mental health and movement activities. Both have helped expose the roles of psychological processes and materials in the formation of belonging and trust in health contexts. With regard to mental health, Tucker (2017), for example, considers the ways in which intensive psychological connections are established between human bodies and material objects within the rural life spaces that constitute sufferers' everyday experiences. Tucker argues that to belong thus comes from, and becomes, a topological physical and mental act of connecting with other humans and non-humans. Along the same lines Tucker (2010b) talks about how life spaces become known as territories through situated relational interactions between particular bodies and objects that become dominant and recognized. The author's case study of a mental health day centre illustrates this. Here, a user describes a place, for him, emerging through the functional activities he undertakes; in theoretical terms, what might be thought of as the creation of expressive, visible and unique 'blocks' of space-time:

> I get involved here in the games room; pool, snooker and err having a chat and a cup of tea with my mates in the Smoke Room. Playing music, which I like to do, Sixties stuff. That's about it really. It's so nice coming

here, you know. I have made a lot of friends here and all that and have a good laugh and everything.

(2010b:439)

With regard to movement activities, Barratt (2012) explores the intimate physical, sensory and psychological relationalities in recreational climbing, specifically between climbing bodies and minds, equipment ('kit') and rock faces. Indeed, Barratt agues, for example, that kit goes way beyond being something of mere use value, being something that is known sensorially, providing enhanced physical potential. One of his respondents comments specifically on ice climbing:

... [E]very [axe] placement you get this lovely squeak squechy scewtchy noise – you can hear and feel that it's secure. A brittle clink or clank and it might dinner plate [shatter]. It's the riskiest but most rewarding type of climbing. You're literally connected to your kit – you feel bionic.

(2012:47)

I often place [protection] for psychological protection. I know if I fall on it, it will rip but what can you do? I have a mental trick though, when I clip my rope I let the gate on the karabiner click, as hard, and as loud, as possible, and that is the mental trigger, that says, I'm safe, my gear is working, climb on; it's scary but it works!

(2012:50)

Finally, although relational materialist thinking has been a fundamental tenet of NRT and more-than-representational health geographies, it is also important to other theoretical traditions. Notably, for example, it has figured heavily in Jackson and Neely's (2015) recent call for a 'political ecology of health' that, the authors feel, could look beyond the body, giving voice to more-than-human things in both disease-making and health-materializing processes.

Onflow

Collectively non-representational theorists regard the world, with all its actors and actants, as continually in the process of becoming. As human subjects, the world's unfolding is at the edge of our awareness, occurring at precognitive level – although we, as human subjects, are wholly bound in it.

(Boyd, 2017:5)

There have long been calls in human geography to consider, more critically and directly, the nature, character and experience of the 'flows' of the social world across space. This is mainly a way of escaping the production of what are regarded as rather static snapshot representations of the world by much contemporary social scientific inquiry, and empirically of accounting for the nature and increasing prominence of movement in an ever globalizing and technologically driven world (Shields, 1997). NRT fully embraces this view (see Chapter 6) but also extends thinking, its idea of 'onflow' – as the word itself somewhat self-describes – incorporating not only flow, but specifically acknowledging a particular moment in that flow: the current, continuing moment at the very frontier of life, at the forward-moving leading edge of space-time (also see Pred, 2005).

There are broadly three reasons why the moment of onflow is attributed particular prominence in NRT. First is its centrality to all life; the fact that onflow happens all the time and everywhere. It being the only place where human beings physically reside and experience first-hand. It being the place where humankind participates directly in the progression of the rest of the universe, and it being the place where all life is brand new. Second is the potential onflow holds. Indeed, onflow speaks to the idea of the emergence of life within its own plane of imminence; life being open, changing and developing right at the point of its appearance. Hence onflow is recognized as a moving edge of sorts, a moving pivot that makes the world constantly unpredictable and subject to immediate change, but also unlocks its unlimited creative potential – in it lying a great deal of the human spirit. Third, and perhaps most fundamentally, is the way in which onflow helps us rethink what 'life' is, helping propose a more-than-biological and active definition that includes many immediate, physical components and characteristics; hence life being all this unfolding together; 'the life' around us, familiar to us all.

Two debates have emerged in NRT with regard to the nature of onflow. The first explores the relationship between the physical and cognitive at the moment, and in the motion, of onflow. The process philosophy and metaphysics of Whitehead, Bergson, Pred, Deleuze and others have proved to be particularly insightful here, helping scholars to unpack the ways in which onflow happens and is registered as an unbroken unified stream of physical and experiential becoming (Pred, 2005). As Boyd (2017) suggests, Deleuze's idea 'the fold' has been useful. Formulated by him largely as a reaction to academic accounts that separate the external from the internal (e.g. surface/appearance vs depth/essence or matter vs consciousness), it argues that the two are actually folded onto each other, thus are of one another, part of one textured happening. This understanding, Boyd notes, has consequences for how academics think of time unravelling; not as something 'else' humans observe, act into and contribute to, but instead as something 'of humans' and their lives.

The second debate is related, concerned with the basic apprehension of space-time at the point of onflow. As Boyd (2017) argues, for Whitehead 'actuality' is comprised of numerous momentary physical entities and events (actual occasions); humans perceiving with their senses, strings of these actual occasions which, flowing from one to the other, appear continuous to them. Moreover, as McHugh (2009) points out, onflow is not mathematically/regularly spaced and timed, but to humans is relationally determined and experienced. Indeed, for McHugh, Bergson's 'duration' captures this thinking. Here duration – or 'real time' – is a duality in movement whereby at each instant the present is simultaneously dilated toward the past and yet contracted toward the future (i.e. our pasts are brought virtually along by us, whilst our futures appear to be rapidly approaching), this consciousness being part of the power/force behind life's ceaseless creation (Bergson's *élan vital*). As Lawler and Leonard (2016) note, in his work Bergson provides various images to help scholars think about duration and its qualitative qualities. One of these images is that of two spools with a tape running between them, its unwinding from one representing our ageing (and our future lives knowingly becoming ever shorter), and its winding onto the other representing our pasts (and our conscious memory bank of them becoming ever larger). According to Bergson, duration resembles this overall image which presents a continuity of experiences without involving static juxtaposition; our current progress and situation in relation to both our past and future. Notably, one consequence of this understanding is that no two successive moments can be identical (because the second will always contain the memory left by the first), and hence every moment in time – in onflow – is always new and absolutely unique.

Onflowing health geographies

Health geographies might not be concerned with elaborating or testing the aforementioned theoretical ideas on onflow directly, or even regularly naming it as a specific concept. Nevertheless, onflow is certainly present and animated in the findings of a number of empirical studies that convey immediate unfoldings related to health and wellbeing to readers.

In studies of health care, researchers have described the institutional creation of onflow through particular interventions that set a certain pace to clinical settings. Solomon's (2011) study of affective relations in Indian medical tourism, for example, describes television sets located throughout a hospital continually playing messages – presented like glossy infomercials on a repeat loop – about the 'loving', 'healing' and 'professional' nature of the care available at that particular institution. Solomon conveys a situation in which one patient used these to get to sleep. Having heard the messages so many times, the exact meaning of the words had begun to take second place to prosody (their overall sound and sensory contribution). They assumed a role in creating a familiar, reassuring rhythmic background hum. An American-accented female voice spoke over soaring uplifting song:

> Reach out and touch…
> It's simply magical, knowing you have the power to heal…
> Every day, every night…
> Saving lives by the minute …touching lives! (2011:112)

The theme here of artificially produced or manipulated onflow, and the role of sound in particular, is continued in recent studies of popular music and health. Andrews (2014c), for example, considers the structural features that make the band Daft Punk's song 'Get Lucky', the feel good tune of 2013, often upping the energy where and when it is played. Andrews observes that the sound energy of Get Lucky comes through in particular technically manipulated forms. It runs at a tempo of 116 beats per minute with a chord structure (Bm7-D-F#m7-E) that does not vacillate. Andrews claims these facets give the song a consistency, and drive it straight through its four minutes and eight seconds with minimal change. He continues:

> From the outset in Get Lucky two introductory bars, each with four riffs, present a simple rhythm that showcases the entire song to come. All instruments – guitar, bass, drums and various keyboards and electronics – enter the fray at once, yet it is the bars played on a lightly effected Stratocaster guitar, that immediately stand out. The riff, constituted of eight or so quick strums, is immediately infectious, and begs you to move with it, to be part of it. Rogers' crisp 'funky' movements and each of his strums, is responsible for the song's immediate momentum. It is a momentum, like with much classic funk, that is constituted of hundreds of tiny moments, with just a micro-second of anticipation and expectation for the next moment to come.
>
> (2014c:10)

Andrews goes on to describe how the immediate 'catchiness' of Get Lucky results from refrains and ongoing unresolved harmonic tension. With regard to the refrains, he notes that its riff is repeated one hundred times in three very close varieties (in four chord loops), and its lyrics – such as 'I'm up all night to get lucky' – also repeat. With regard to harmonic tension, he notes that the song does not contain or rely on a 'build', but instead generates an addictive positive energy by simply never settling to a 'home chord'. This sets up an underlying anticipation which is never resolved or released, just recycled and set back in motion. Andrews observes that because listeners have never reached the song's destination (because one never emerges from the tension), they are forever taken on its journey. Notably, Andrews describes how in Get Lucky the artist is explicitly commenting, through lyrics, on events and situations which have wellbeing implications (namely on young people having fun on nights out and entering casual sexual relationships). The lyrics alone might encourage conscious reflection on these events and situations. However, the structures of the accompanying instrumentation do something complementary. They create a forward moving onflow of energy and mood that strongly evoke the very same events and situations. Thus they might enroll bodily participation in that moment, starting and/or intensifying conscious reflection.

Onflow is, of course, just as relevant and present in more organic contexts. Andrews et al. (2014), for example, consider the emergence of wellbeing as an affective feeling-state in everyday life. Their field notes, taken in a lakeshore park on a public holiday, highlight the importance and combination of heat, light, colour, movement and sound. In the extract below, the researcher is sitting soaking in the warm afternoon sun, his legs stretched out in front of him. He is looking through a heat haze watching children play in a splash pad. The notes describe the scene unfolding and his feeling of wellbeing gradually emerging:

> The kids' laughs and shouts are accompanied by the constant pitter patter of the water falling from high parts of the structure and onto the floor – rat tat tat, rat tat tat, rat tat tat. I gradually look to the left. Hundreds more people are in the park, together appearing like a single entity. People sit under sunshades, others walk in all directions. This is their movement. Hundreds of conversations, making happy waves of fragmented voices; a few random words, parts of words, letters become clear in the waves. A band plays over them. This is their sound. A sea of red and white hats, t-shirts, painted faces. This is their look. Feeling the heat and energy, I look back towards the splash pad. The sound of the music and chattering public now joins the pitter patter of water. My son comes directly into view in front of me splashing in the water, slap slap. A red balloon floats to my feet and comes to a rest. Feeling energised by it all – in my legs, arms and mind – I get up. I have an even bigger smile :)
>
> (2014:215)

Research also considers the interplays of the natural, physical and cognitive in onflow, and their relationalities within the assemblages involved; walking often being the particular activity used to elaborate these. Wylie (2005), for example, talks about the therapeutic experience of his walking fieldwork in Devon, UK:

> A walker is poised between the country ahead and the country behind, between one step and the next, epiphany and penumbra, he or she is, in other words, spectral; between there and not-there, perpetually caught in an apparitional process of arriving/departing.
>
> (2005:237)

Likewise, Macpherson (2009a), considers how landscape becomes present in the intercorporeal experiences of walkers with visual impairments during their walking group activities. She argues that it does so through their movement through landscape which gradually opens up in many other-than-visual ways. This being a coming together of emerging physical terrain, bodily and

Figure 3.1 Vitality

interbody efforts, sensory experiences, visualization, imagination, personal and known histories and conversation.

Vitality

Across the sciences the term 'vitality' is often used quite loosely to denote fundamental diversity and strength within a group/set (perhaps, for example, facets in a biological population). As mentioned in Chapter 1, however, in NRT and related modes of thinking, vitality is about the aliveness and buoyancy of the living and non-living worlds and their collective potential. With regard to the living, direction is gained from vitalist philosophy which recognizes the exceptional qualities possessed by all living things; their essential spark and energy, and that they constantly move, change and evolve (Philo, 2007; Greenhough, 2010). With regard to the non-living, direction is gained from new materialist thinking, particularly on the subject of 'vibrant matter' (Bennett, 2009), which moves beyond physics' definitions of vibration (the expenditure of energy through oscillations around an equilibrium centre), to describe the capacity of things – from core materials to more complex objects – to act as quasi-agents with their own tendencies, trajectories and forces. Both of these lines of thinking recognize that when encounters happen between the living and non-living, an energetic animation results; a collective materiality that gives life a range of qualities including its richness, diversity and spirit, its endless capacity to develop on its own impulses, its self-generating continuance and purpose, yet also its instability, irrationality and unpredictability. Vitality is of course tightly bound with the idea of affect (explored fully in the next chapter), and it is not entirely possible to separate the two. One might thus think of vitality as resulting from a process of energy enhancement made possible from affective interactions between vital bodies and vibrant objects.

In sum, as an intellectual idea, vitality provides a relief from deterministic academic thinking in the social sciences by demanding a re-evaluation of how we understand events and things, not as definable by their properties but as always animated, energetic, in flux, both creating and open to new temporary relations (Greenhough, 2010). So, as Greenhough describes, as much as vitality is reflective of certain qualities life possesses, it is just as much a way of acknowledging the limits of our current understandings, begging that, as researchers, we attend to the make-up, character and push of the world.

Vital and vibrant health geographies

Three areas of empirical interest constitute a good proportion of the engagement with vitality in health geography: CAM, blue spaces and what might be loosely termed working spaces. Within each area, energy – either new energy, increased energy or renewed energy – is a common theme. With regard to CAM, as noted in the Preface, Philo et al. (2015) explore the energetic uplift yoga and meditation enact through the interaction of body–mind practices and objects. The authors note that one of their participants/ diarists walks briskly to work through a park, meditating a little under the trees. She calls this regular event her 'meditative walk', that creates positive energy:

> {T}he benefits [are] not just physical … but also helping with general mental health … a positive energy level. … On a physical level, cardiovascular, … [and] the mental aspect of extensive enlivened energy as well as a sense of calm, and when I can do the meditation it also helps centre oneself.
>
> (2015:36)

Further, the same participant/diarist also comments on her more formal after-work yoga class, undertaken at a specific venue:

> So I think it's very important to have that space to go …, because the energy in a room makes such a difference, you know. Just the fact that there's all those other people and that you've all come there together, and you are all there for … similar reasons. And there's just an amazing energy in a room and particularly in the self-practice class because all you can hear is just people breathing. And everyone is just totally focussed on what they are doing and, yeah, it's just energising and uplifting; just being there makes a difference.
>
> (2015:37)

In another CAM-focused inquiry, Andrews (2004) describes the practices therapists employ that concentrate their clients' minds on specific body parts and on their 'metaphysical energy'. A shiatsu therapist in this study, for example, commented:

> I talk to them and get them to be attentive to places of touch. I encourage them to try and feel certain places within their body ... You try and make the client see his or her body from within for them to connect with their own healing power and feel the energy around their body. They concentrate on the abdomen for deeper breathing, the legs for more grounding earth energy, chest to contact feelings and emotions, back for letting go of tensions.
>
> (2004:312)

Similarly, Paterson (2005) describes how reiki deals with the body's 'universal life force' (*chi* or *ki*) passed on through therapeutic touch which opens up another body's energetic pathways, realizing its potential.

With regard to 'blue spaces' – notably a more general emerging interest in the sub-discipline (see Foley and Kistemann, 2015) – scholars have discussed the vibrancy of water, the vital qualities of the biological life it hosts and the overall vitality of the encounters that occur with and within it. Foley's (2011) aforementioned study of the holy well describes these as a therapeutic assemblages comprised of diverse mixes of bodies, objects, symbols and behaviours. Healing energy results from the moment of contact between bodies and water, which thereafter provides visitors with renewed strength in their daily lives. Foley states:

> [E]nergy... is a key ingredient in this spatial stew... material settings and bodies hold within them a range of potential energies, those energies have symbolic healing associations and it is through energetic performances of health in place that productive aspects of the assemblage are expressed.
>
> (2011:472)

Meanwhile other studies of blue space have focused on activities that are a little more mainstream. Foley (2015), for example, considers sea bathing as a therapeutic practice; moving swimming bodies in motion with a 'natural' body of water that, rather than being an inert suspending environment, has a sensual energizing quality of its own. Moreover, Evers (2009) describes surfing and the bonding it involves between the moving body and the moving energy of the sea. What to do when catching a wave, the author argues, becomes intuitive through an ongoing embodied learning process with the lively ocean; such decisions being made through an emotional and affective backdrop of adrenalin, fear and excitement, all related to potential outcomes. Finally, Humberstone (2011) considers the 'lifeworld' of windsurfing, of being in and moving with nature (across water with the wind). The author describes this as a healing and spiritual experience born in a 'kinetic empathy'. A section from their participatory field notes reads:

> I sense the wind shifts in strength and direction and move my body in anticipation to the wind and the waves. I feel the power of the wind and the ability of my body to work with the wind and the waves. The delight and sensation when surfing down a small wave with the sail beautifully balanced by the wind.
>
> (2011:502)

With regard to working spaces, certain literature thinks of these quite broadly as ranges of services which, through the energy they possess, enhance the wellbeing of users who move between them (see Walton, 2014). Most studies are however concerned with quite specific types of working spaces. For example, Kraftl and Horton's (2007) study of 'the health event' (a small conference where the findings of a study on young peoples' health needs were shared with those who had participated in it) observes how the great deal of energy and effort expended in planning the event resurfaced in the event proper on the day it was held. The authors note how this was assisted by formal structured activities including the introduction of specific objects (such as video, audio recording and a graffiti wall) and lively practices (such as role playing, brainstorming and group breakout discussions). They note how the result was an affectively produced vitality that cannot easily be measured or fully explained:

> In the case of our event, this [the non-representable ways in which events occur] meant attending to the sheer work without which the event could not have happened, and which was largely responsible for imbuing the event with its 'participatory' vitality. The event, and the knowledges and 'usefulness' that it produced, emerged in more and other ways than flip-chart lists of recommendations, or formal reports, might initially indicate.
>
> (2007:1016)

It should also be recognised that an opposite process can equally occur, whereby when objects or bodies are removed from a particular assemblage, the place loses its affective qualities and vitality. In Andrews et al.'s (2014) aforementioned study of everyday places of wellbeing, Sam, a co-author, reflects on her workplace after closing (typically a busy and 'happening' restaurant with bodies moving, food cooking and being served, ambient lighting, cool music and lively conversation):

> 'Is there something I can help you with?' I politely ask a man who is wandering around my workplace late at night after it had closed. 'No, I'm just looking around', he replies. 'I've never seen this place with the lights actually on before. It looks so different!' I join him in looking around,

slowly surveying the place in a slow moving arch of my head. I am struck by his observation and how right he is. The harsh light and absence of people draws my attention to the empty perfectly set tables and chairs, so stationary, so ordered. Without the smell of food, people conversing, low light, ambient music, and being constantly on the move between tables, it feels awkward and empty; a shell of a place.

(2014:217)

Virtuality and multiplicity

NRT appreciates the virtuality and multiplicity of space-time. At its most theoretical, this has involved a Deleuzian reading of 'the virtual' as something other than 'the actual' (i.e. observable properties or traits) yet still real: either surface forces and affects produced by actual events, or a form of potential

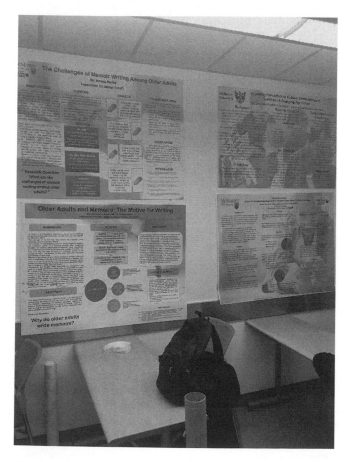

Figure 3.2 Virtuality and multiplicity

whereby combinations of forces and affects push forward and are felt, but are not necessarily actualized (Cadman, 2009; Duff, 2014a). As Evans (2010) suggests, it is the potential of an actual outcome that makes the virtual important in life, particularly in terms of how it promotes pre-emptive human action. She argues that, more often than not, pre-emptive human action is not the result of logical consideration of evidence on what might happen in future, but instead is legitimized by the reproduction of 'affective facts' which make potential futures resonate and felt in the present (i.e. what something might feel like were it to occur).

In speaking to the future, the virtual can draw on the past. McHugh (2009), in a theoretical study of ageing, memory and landscape, draws on the work of Massumi, Deleuze and others, arguing that the human passage through time and space is not a succession of isolated states or snapshots. Instead, he posits, as suggested earlier, our experience is that of unbroken duration, of a dynamic continuation of movement and sensation. He argues that the past is not dead simply because it is not currently happening. The past exists in a virtual dimension, its affects and other feelings being reactivated by memory images (thus the past (virtual) and present (actual) are not completely successive/concurrent and they can be lived through each other). Indeed, for McHugh, certain events and experiences can leave affective memories which can re-emerge as current sensory experiences re-evoking the events, rather like hauntings from the past (Amin, 2004; Malpas, 2012).

A Deleuzian reading of 'a multiplicity' is of a unique and new complex entity that originates in interactions between simpler/basic elements; an entity that is always in flux and open to inputs/modifications and that can cover space and time (such as age, gender, class). Given the existence of virtual forces, different multiplicities arise simultaneously in places imparting their qualities depending on ones viewpoint. Even when these multiplicities act together the result is still often formless and groundless. As Bergson (1889) points out in his classic work *Time and Free Will*, a multiplicity and temporal heterogeneity exists whereby in the present several conscious states permeate one another but combine into a rich conscious whole.

More practically and less abstractly, however, virtuality (read in popular/dictionary terms as 'almost but not actual') and multiplicity (read in popular/dictionary terms as 'a state of being multiple') can also speak straightforwardly to the way that space-time is ruptured (non-fixed and non-linear) and the relatedness of life across these ruptures. As Andrews (2014a) describes, this might be realized, for example, in things (and places) that are 'real' to people in terms of a sensory experience, yet are not entirely physical or present (for example seeing a place on a website or through a visualization technique) and/or multiple happenings that are related and similar in terms of sensory experience, but emerge and co-exist in different places (for example, one form of practice rolling out simultaneously across different settings), and/or different happenings in different places being experienced in a sensory sense in a single space-time (for example, communicating through teleconferencing).

Thrift (2011) argues that in previous historical eras, geographers and others were able to place cartographic grids over the world, whilst later they were able to track movement over it. But now, with modern lives being lived through omnipresent technologies – and in the security–entertainment complex Thrift mentions – a new era of research might have to exist, less to do with any form of calculable coordinates, and more to do with the gradients of resistance that create the world. For Thrift, what counts as existence and non-existence is itself even up for debate, he proposing that multiple ontologies might co-exist in, and constitute, our faster and ever more complex world; different spheres of existence situated side by side, every one incomplete but completing the whole. As we shall see in the following sections, in health geography virtuality and multiplicity arise in two broad fields of research interest, one focused on practices and care, the other on everyday encounters and life.

Virtuality and multiplicity in health geographies: practices and care

A popular field of empirical inquiry that showcases the aforementioned theoretical ideas is CAM. With regard to the virtual, and its potential, Andrews' (2003) study of older people's consumption of CAM talks about the 'Totnes phenomenon'; Totnes being a small rural town in the UK with a large number of CAM clinics for its size, and that possesses notoriety for its highly spiritual atmosphere. Andrews claims that this atmosphere is produced, not only by the presence of these clinics and by the nature of the local shops and services available in the town (including those selling 'alternative' clothing, artefacts and music), but also by the everyday attitudes, pastimes and actions of residents and visitors. In addition to the utility derived from the services on offer, this spiritual atmosphere was known to be part of the overall attraction of the town as a destination for CAM users, they being drawn in by the overall feelings they expect to experience, and do experience, once they arrive; this being akin to a virtual force – a felt potential. A respondent states:

> Therapies are only part of the total Totnes thing. You've seen what it's like! The school children busk on the street, lots of people with funny clothes, all these rare shops. It makes a nice atmosphere, but it's hard for me to summarise it all for you. You just have to experience it.
>
> (2003:342)

Moreover, the same respondents talked about affective experiences of their specific CAM therapies being 'reactivated' as they went about their daily lives thereafter, often beyond Totnes (for example experiencing moments of mindfulness or calm). The same kind of affective reactivation occurs for the users of yoga and meditation in Philo et al.'s (2015) recent study of those particular holistic modalities; again something akin to a virtual force seemingly being at play.

With regard to the aforementioned more popular/dictionary readings of 'virtual', much research connects here, again a substantial proportion being on CAM, which, as Doel and Segrott (2004) suggest, is a type of medicine and care which has no one universal form, 'taking place' in a multiplicity of registers and domains (from therapies to therapeutic products and media, from the physical to the mental). Indeed, theoretically, Andrews (2004) argues that CAM challenges the necessity of physical co-presence in the production of therapeutic experiences through its use of visualization techniques. For example, a therapist in this study used re-visualization of familiar places to help their client imagine a life without fear of spiders. They state:

> I get her to envisage a scenario where they are not frightened of spiders, where they can walk around, live with them and feel safe. It involves places where they live and work.
>
> (2004:312)

> It might go something like… 'imagine sitting in your living room at the end of a long day relaxing and watching television, and the spider is there, on the wall, minding its own business, living it's own life. You feel for it, it being alone and interested only in survival. But he's fine now, you're both fine'. You can enjoy your evening.
>
> (2004: unpublished data)

Andrews goes on to explain how the places suggested by therapists and imagined by clients may be 'real' (in the sense that they may physically exist in other times and places) but how equally they may be virtual (effectively created by therapists and existing solely in the mind). A therapist discusses this issue:

> Much depends on the client and their situation. Real places are appropriate to use at times whilst at others I make them up. I do not really think about whether they actually exist, and sometimes there is a grey area when you are suggesting places and the client may or may not picture it as somewhere that they have actually been or that they know.
>
> (2004:312)

In terms of techniques, Andrews describes how another therapist in his study evoked sensory experiences through visualization practices but, creating connections and a multiplicity, matches her voice and delivery in the clinic to the places evoked:

> They have to be quiet places. They are very peaceful and in the presence of nature, like near running water, at the seaside, on a beach. Warm

places are good because warm and water induces relaxation … I have to suggest these places in a calm and relaxed manor, in tune with what I am describing.

(2004:313)

Similarly, but in the context of conventional health care, Andrews and Shaw (2010) investigate visualization as an informal professional practice that doctors and nurses employ when encountering sufferers of needle phobia, to take them mentally away from the traumatic situation they are facing. A respondent in their study recalls a specific situation that had recently unfolded:

I had a young lady who had a tap done. Of course they were pushing on her chest, she is holding my hand for grim death and, you know, so I am talking to her. It's not something that you necessarily plan sometimes. I was trying to keep her calm so I asked 'where would you like to be right now?' and she said, 'Oh God, I would like to be…'. So we started talking about a beach in Barbados. It was just to take her away from that situation. You can't physically transport her, but you can take her away in her mind, so she's not focusing on pain.

(2010:1806)

Indeed, in these cases the virtual involves the bringing in of other places, present in non-physical forms, yet acting into the moment as an active part of the therapeutic assemblage. This is the mental bridging of, or movement between, 'there and here' and 'then and now' for therapeutic ends.

Virtuality and multiplicity in health geographies: everyday encounters and life

Away from specific care and therapy contexts, studies have shown how the virtual arises in, and facilitates, everyday experiences of wellbeing. Evans (2014), for example, explores how these experiences might be rooted in technology. Specifically, he studies the affective therapeutic qualities of electronic ambient music. Examining the works of Brian Eno, he talks about techniques the artist pioneered such as tracking, mixing and signal processing, all undertaken to create a sense of spaciousness and otherworldly atmospherics that blur the distinction between noise and music. In addition to his technical analysis, Evans' participant field notes, changing in font, reverberate an affective experience of listening to Eno's 'On Land' whilst lying on his office couch. His is a journey of the mind travelling through the locations painted by the sound, but one not divorced from his immediate physical context. Evans' body and mind interplay, interconnecting the various levels of his existence in that moment:

My journey begins at **Lizard Point**. My attention is immediately drawn to the murky drone – my body settles into the couch. New sounds continually enter the scene. Ducks call nearby. Deep crashes of thunder (or bombs?) in the distance. Its dark. Its wet; water drips nearby. I notice that my breathing has slowed and steadied, in... and out, in... and out, in... and out. **The Lost Day**. The drone is more intense and is now offset with the repetitive banging of metal in the distance. My body's position has not changed, legs still crossed, lying perfectly still. My arms and legs feel completely relaxed. Breathing still the same in.. and out, in... and out, in... and out. My mind is attentive and tracking the sounds near and far. **Tal Coat** begins with an electronic buzzzzzzzzzz like a transformer box. I hear water bubbling below me. Abrasive waves of white noise swell. The noise of the song is intensifying. My body is still relaxed but a dash of excitement runs up and down my back. A light piano phrase drenched in reverberation followed by a three note base riff eases the tension. My body still has not moved and my eyes look across the office, focused on nothing really. s t a r i n g.

(2014:183)

Exploring the idea of multiplicity in health experiences further, in terms of the levels and qualities involved, Wylie (2009) studies memorial benches on coastal paths. A sense of absence, apartness and loss experienced at them, he argues, is refracted by the external contemplation of a scene and its revealing textures, colours, distances. These, he posits, are landscapes framed a certain way by the positioning of the bench, but that themselves change with light and the time of day to reveal uniquely on each occasion. Wylie reflects specifically on drifting, aerial, untethered, passing encounters:

[A] series of tensions between watcher and watched, interior and exterior, the invisible and the visible, are set in motion. You might seem to feel (for a minute, as I'd done) the living presence of the coastline, the sea and the sky, with this experienced as a sort of sublimely de-personalising tuning-into or becoming-with: phenomenal coincidence of self and landscape. Or else (remembering the dead), the benches might be for you a sort of meditative and reflective resource, a place where things could be put in perspective, and look out there to the far horizon, now it's truly a vanishing point.

(2009:278)

As Wylie posits, it is the apartness that facilitates these therapeutic moments of conscious contemplation. Indeed, love, he argues, is felt most acutely in all our lives when we are apart from what we love (when there is a gap/fracture); the benches and scenes replicating this feeling somewhat. Wylie's acknowledgement of the physical distance from particular elements of the landscape contrasts with much research in health geography which tends to emphasize bodily presence set deeply within therapeutic landscapes and immediate 'close-up' sensory experiences.

Hope in potential and becoming

> While there is life, there is hope
>
> (Stephen Hawking, 2006)

As suggested in Chapter 1 and at other points in the book thus far, NRT emphasizes the idea of immanence. Life not as the result of structural impositions (such as codes or laws) of external/higher transcendental phenomena (such as god, reason or rules ascribed to nature), but as a self-generating and self-organizing process encompassing and manifested in the actual material world itself; life being the drawing of matter together in its own field of becoming or plane of immanence (Duff, 2014a). In this emphasis and understanding lies a particular version of hope; hope not as traditionally or popularly understood as in some hypothetical or utopian endpoint or something aimed at, but hope as part of life emerging; its potential and its movement. In terms of potential, hope arising because of the unfinished nature of all situations and because life always possesses the possibility of something else (as Anderson, 2006. describes, life with a partial and escapable present, and future that is open to difference). In terms of movement, hope generated and felt by affective and other encounters (hope feeling like a body's enhanced capacity to act and this acting itself (Anderson, 2006; Duff, 2014a)). Hope then can fade if the possibility of change/movement is reduced, disappointing and depleting the experience of the present space-time. On the other hand, hope can build through a realization that new possibilities for change/movement have arisen, which can reanimate the present space-time (i.e. which is life-enhancing in itself). Notably hope has two places in NRT. Not only, as just described, is it in the unfolding of non-representational events but, as we shall see in Chapter 8, it also resides in the methodological practice of NRT; in the ways that NRT seeks to 'act into' life, helping chosen aspects of the world to speak back.

Hopeful health geographies

Traditional health geographies are certainly not bereft of hope. Popularly understood as an endpoint or ideal, hopes permeate all fields and debates in some way, particularly because health often involves life and death, happiness, comfort and quality of life. As individuals we hope for good health for ourselves, for our loved ones and for our communities (whilst our hopes might be realistic, idealistic, unrealistic or false, and we might need hope, lack hope or have our hopes realized or dashed). Most often in health geography hope is an undercurrent and/or occasional point in research, existing or made with varying strength depending on the particular study. Moreover, hope exists in different places in different theoretical orientations of the sub-discipline. In spatial science hope might lie in exposing a specific distance to a health service that needs to be reduced or in an intervention to reduce the spread of disease.

In political economy hope might lie in exposing the successes or shortfalls of particular policy, funding or financing scenarios. In social constructionism hope might lie in exposing how bodies or health or disease are being represented by officialdom. Otherwise, as researchers we might hope for the betterment of things and situations (a better design, better service, better policy), of people (their food security, access to services, good health and wellbeing, longevity), and also for the betterment of research and our sub-discipline/ field (often through changes in direction – method, theory, empirics).

In contrast, in line with the earlier discussion, in more-than-representational health geographies hope lies in unfinished business, in potential and movement. Anderson (2002), for example, in his study of the practices of listening to recorded music – to 'feel better' and escape from depression – suggests that hope might involve two forms/enactings of affect: a virtual evoking of the affective idea and possibility of how something else might feel, and positive physical affects that might help one forget worse affective and other predicaments. In this extract, a participant in Anderson's study finds escape through movements that help her forget the circumstances surrounding her partner's son who was suffering from clinical depression:

> Andy had gone back to the hospital yeah … in a temper. [I] put Fraiser to bed and I came downstairs and I would normally do a bit of housework, might watch TV … or I might read my book … and I thought, and it was not because of this, but I just thought … I want to put some music on, LOUD. So I put Ska music on and I danced, and danced, and danced … and it were BRILLIANT, absolutely brilliant … I managed to … switch off that somebody might come back thing, and … kept an eye on the door … you know … I had a great, great time … and Rat Race and there were some Selecta ones on there as well … I think … and then I got a Beats CD and I put that on and I danced to a bit of that, it's funny after that … I stopped … and thought … GOD you're 35 stop this, I don't know where it came from, just a little thought in my head, put the brakes on, grow up, turn it down … you know … get the washing up done … and I screeched to a halt with it and felt really silly for a bit afterwards. I DON'T NOW … but in that moment I did I just danced, danced me heart out and it all lifted … just for a bit … it's a cliche but I WAS somewhere else.
>
> (2002:211)

Thorpe and Rinehart (2010), on the other hand, consider how the movement-activity of alternative/extreme sport enacts a radical politics of hope. Specifically, drawing on Thrift's (2004b, 2008) thesis, they argue that we are entering a societal stage less attached to political parties and organized politics and more directly involved in action and alternative ways of expressing in person and through technology. Indeed, they argue that engaging with the typical post-liberal politics of 'evidence' can be depressing. Far more attractive, interesting, and ultimately more fruitful for many, being a reactionary

counter-politics that lies outside traditional political structures. Here immediacy, feeling and performance count, the authors commenting:

> There seems to be a movement to new forms of sympathy – new affec-
> tive recognitions, new psychic opportunity structures, untoward reani-
> mations, call them what you like – forms of sympathy which are more
> than just a selective cultural performance and which allow different, more
> expansive political forms to be built... this turn to alternative ways of
> thinking and being is particularly true among younger generations who
> appear to be less willing than their parents and grandparents to channel
> their political energies through traditional agencies exemplified by parties
> and churches, and more likely to express themselves through a variety of
> ad hoc, contextual, and specific activities of choice, increasingly via new
> social movements, internet activism, and transnational policy networks.
>
> (2010:1278–1280)

This type of change can be seen, for example, in the ways in which health
causes – from the singular patient to system level concerns – get pushed and circulated on social media, and how they often involve protest and direct action.

In sum, the conceptualizations outlined in this chapter lay an important
basis for more-than-representational health geographies, and provide ideas
and tools to animate the active worlds of health. There lies great potential for
each to be further explored and used in new empirical contexts.

4   A mode of health transmission
Affective health geographies

This chapter introduces and explains 'affect' which, in addition to being the main working concept of NRT, is now so popular across human geography as a whole – part of the wider adoption of post-humanist, new materialist approaches – that it could be argued that the parent discipline has undergone somewhat of an 'affective turn' in recent years. Following this, attention is paid in the chapter to how affect arises in more-than-representational health geographies, and how it adds to their explanations of health within four empirical fields where it appears most regularly: health care and health research, coping with chronic conditions, health-harming and joyful behaviours, and the politics and wellness in movement activities.

What is affect?

As Thrift (2008) notes, affect has a diverse philosophical precedent and grounding in the work of such eminent scholars as Plato, Kant and Rousseau, but, importantly, is not held to any one common understanding. A number of explanations exist that span hundreds of years of academic thought, including phenomenological and social interactionist (emphasizing embodied practices that create visible actions), psychoanalytical (emphasizing practices that emerge from and as human drives), Darwinian (emphasizing expressions of emotions that are similar across species) and contemporary psychological (emphasizing experience and display of human emotion). It is however a fifth explanation, a specifically naturalistic one, that has been most influential in the social sciences in recent years including in NRT. Originating in Spinoza's *Ethics* and his early philosophical reasoning of mind, body and nature, the naturalistic explanation was later developed and articulated more extensively in Deleuzian critical theory (Deleuze, 1988, 1995). Here affect is built on some fundamental academic rethinking of the nature of the human body. As suggested in Chapter 2, Deleuzian philosophy asks not what the body is (such as how it is composed and works), but what it can do (its potential) and what it does (the results of its expenditures of energy) both individually and 'transpersonally' (produced through, circulated amongst and shared between bodies beyond the limits of personal identity) (Duff, 2014a). As Duff argues,

Figure 4.1 Affective

from this perspective, the body is both complex and open, meaning that it is composed of numerous parts each of which – if they are fully operational – are permanently amiable to relations with their surroundings. This creates a situation whereby the body's capacity to be influenced is not single or fixed. Indeed, as Duff suggests, the body's facets – such as expressions, habits, movement forms, ways of speaking – are folded into the body as transmitters or orientations; all of the body's key developmental domains and capacities – social, cognitive, emotional, physical and moral – being acquired partly in its engagements with the world.

With this version of the body as a starting point, affect then can be understood as a transitioning of the body and the process whereby it is affected by other bodies, modifies and affects further bodies (see Thrift, 2004b, 2008; Anderson, 2006); the transition from one experiential state of the body to another encompassing changes in its energy, which is either amplified (positive affection or a joy/*laetitia* in the body) or drained (negative affection or sadness/*tristitia* in the body) (Deleuze and Parnet, 2006; Dawney, 2013). Thus, through providing an energetic uplift or dampening, and impacting human capacity for engagement and involvement in life, affect impacts upon wellbeing primarily in an immediate felt sense, and secondarily through increasing or decreasing involvement in specific activities that might themselves help induce wellbeing (Andrews et al., 2014). Importantly, affects are not fully known to, or reasoned by, individuals as they occur, but instead are experienced less-than-fully consciously by them, revealing on a somatic register as a change in their 'feeling state'. Affect then is a passion of both the body and the mind experienced powerfully if only vaguely realized. Notably, whatever the role of culture in the production of affects (and vice versa), affects contain absolutely no meaning in themselves.

Environmental contexts are critical to the production and experience of affect, hence the attractiveness of the concept to human geographers. As Duff (2014a) suggests, in Deleuzian terms, milieus assembled from diverse human other living and non-human elements help generate affects in particular territories (non-human elements – or objects – contribute affective qualities such as their texture/ feel, movement, sound, light, color, temperature and smell). As a descriptor, then, 'affective environments' might be thought of as the collective expression of affects and their reproduction in places (in other words assemblages where affected bodies interact with other bodies and objects, triggering further affects and these happening). These affective environments transmit – being soaked up and registered by humans both singularly and collectively – thus are experienced as prevailing 'atmosphere' (Massumi, 2002; Thrift, 2004b; Anderson, 2009); a process and experience intimate to the point at which space and affect become simultaneous and co-extensive (Thrift, 2008; Bissell, 2010). Notably, other spatial aspects of affects include their bridging of scales; it being possible, for example, for affects to be generated and felt between two bodies in a room but equally between tens of thousands of bodies – often through technological mediation that transports them over physical distance – at a global level (it also being possible for these scales and circumstances to be related). In sum, then, affective environments can be stable or changing, finite or ongoing, distinct or overlapping, single-scaled or trans-scaled. So, for example, an individual might move between different affective environments, the transition being seamless or abrupt, or one affective environment might be set within a larger one.

Importantly, as Hall and Wilton (2017) suggest, what drives human bodies into affective relations with other human bodies, and more broadly into affective environments, is a basic desire (*cupiditas*) – that is natural but not always consciously realized – to interact. To move beyond the single body and be part of bigger happenings, the bigger physical picture of the world. Offering opportunities to fulfill this desire, affective atmospheres are thus alluring nodes of basic feeling that invite humans in, and once in, to respond (Anderson, 2009). On another level, as suggested in Chapter 3, humans are also driven to moving into or out of particular affective relations and environments due to the power of affect in helping them pre-empt their potential futures (what might happen under particular circumstances and in particular places). Indeed, in this scenario affect arises as a momentary virtual event in the mind and body in the present, allowing us to know how opportunities (or even threats) might feel if they come to be (Anderson, 2007; Evans, 2010).

Notably, as suggested in the Preface, armed with lay versions of these knowledges, state and commercial interests often purposefully create affective environments with specific atmospheres, providing familiar textures to peoples' lives often through, for example, attention to sensory attributes in standardized forms (Thrift, 2004b) (such as the designs and executions of houses, office buildings, restaurants, shops and shopping malls). However, equally affect might be less designed and more organically occurring (such as in a busy city neighbourhood) and/or empowering (such as political potential maximized through demonstrations where bodies come together in a common cause).

There has been some debate in human geography and other disciplines regarding the commonalities, overlaps and relationships between affect and emotion. In terms of conclusions, much has depended on the definition used of the latter. For certain scholars, emotions are known to a greater extent than affects, some involving fully conscious reasoning. Pile (2010), for example, describes a three-layered model that maps the order and relationship between the physical, affective and emotional (see also Thien, 2005; Anderson, 2006). The first layer is the purely physical interactions that occur within assemblages of bodies and objects (here humans, like all objects, are neutral, ahistorical, universal and integrated; Pile, 2010). The second is a less-than-fully conscious affective feeling state; how these physical interactions are tacitly, intuitively picked up, yet not consciously registered or expressed. The third layer is fully consciously felt and known emotion; the way affective feeling-states are later fixed on or compared to established social categories and personal experiences, and expressed. Pile argues that a straightforward one-way movement typically occurs between the first, second and third layers (whether involving fractions of seconds or longer), that the first and second layers can occur without the third, but that no leaps can occur over the second. Whether or not one agrees with this typology, it is fairly safe to suggest that affective environments host and feed into a range of emotions (as Philo et al. 2015. argue, many forms of emotions would not work without affect, and would remain lifeless words and thoughts). Certainly there exists a degree of flexibility in academic/disciplinary traditions, notably the established field of emotional geography incorporating affect into numerous studies.

In sum, as Duff (2014a) argues (and as suggested in Chapter 2), the study of affective relations under NRT is, as an interest, part of ethology – the study of relational capacities, speed, thresholds and transformations. It is certainly radical, not only for engaging with these things in the context of social science research, but also because it moves beyond considering a static lone body: its structure, function and even opinion. At the same time (as suggested in Chapter 1), it is an ethology which acknowledges that there are excessive dimensions to being and the world that remain unqualified and cannot be accurately predicted, observed, determined, measured or even summarized. Indeed, excessive dimensions that are non-representable because they cannot be known by being captured by representational approaches and squeezed into established categories for phenomena (Wylie, 2005; Bissell, 2010). As Deleuze and Guattari (1994) note:

> Affects are no longer [just] feelings or affections; they go beyond the strength of those who undergo them. Sensations, percepts and affects, are beings whose validity lies in themselves and exceeds any lived (1994:164).

Researchers then can only ever observe intent or agency to create affect, or specific components, movements, interactions, sensations or emotions that are only parts of the overall affective process and intensity that itself always remains powerful and palpable yet elusive.

As suggested at the start of this chapter, affect is relatively commonplace across four empirical fields of more-than-representational health geography (focused on health care and health research, coping with chronic conditions, health-harming and joyful behaviours, and the politics and wellness in movement activities). The following sections showcase some of the main studies that help constitute each.

Affective health geographies: health care and health research

Using the idea of affect scholars have described atmospheres associated with particular producer practices and experiences, and consumer attractions and experiences in health care; atmospheres that are critical to the workings and success of various health specialties and sub-sectors. With regard to work life, Ducey (2007) for example, studies allied health care workers in training (nursing assistants, medical technicians and other roles) articulating how their work is 'more than just a job'. By this the author means an energetic passionate impulsive physical engagement between workers' bodies and technical objects, as they exert efforts together in shared directions and create shared momentums on the way to reaching common goals. She argues that, for these workers, this engagement is a positive experience in itself, but it also feeds into the meaning of their work through cultivating an appreciation of the merits of this 'joyful' physical investment and expenditure. Ducey speculates that the affective elements of everyday work life are likely be important to most categories of workers across both health and social care, across both the informal and professional sectors, and across clinical and administrative fields; their unique features being interesting possibilities for future research.

Meanwhile, with regard to the provision of particular consumer experiences, Andrews et al.'s (2013) aforementioned study of complementary and alternative medicine (CAM) located in conventional primary care settings, describes the 'affective possibilities' offered by therapists to potentially entice their clients – who pay for their services largely via direct out-of-pocket payments – back for repeat treatments (and thus simultaneously improve their own health, and the health of the business). These affective possibilities, which are part of the overall intended therapeutic experience, are provided where possible through the purposive design and decoration of practice settings (involving for example, ambient background music, soothing colours and natural materials) and the deployment of particular interpersonal caring practices (including touch and body movement and positioning). A therapist in this study states:

I want them to feel, as soon as they enter, what this place is all about. I want them to feel good but to get a sense of my passion for this medicine, so that they become immersed as I am and develop similar feelings.

(2013:105)

Providing more detail on inter-body communication in CAM, Paterson (2005) describes affective transfers through therapeutic touch in reiki. She comments on the creation of connection and a deeper sense of contact between bodies; of 'feeling with' and of being 'in touch with' another tactile body (that contrasts with procedural or task-orientated touches often part of conventional medical practice). This, she describes, is a form of bodily empathy, working in that moment against the flow of other, often chaotic non-therapeutic sensations in everyday life. It is a transfer of emotional energy that builds bodily capacity to deal with the future.

Elsewhere, along the same lines, in an aforementioned study, Solomon (2011) exposes the 'affective economies' in Indian medical tourism. These are affects intended to contribute to the overall experience of escape, healing and repair, but also designed to facilitate and feed sentiments and perhaps more conscious feelings around medical competency (both ultimately intended to improve consumer experiences and increase future demand). Indeed, Solomon suggests that:

the affect of anticipation guides the journeys of patients seeking what they could not find at home, but it also conditions the handshakes among hospital and government officials that ultimately deem foreigners as possessing more 'return' on investments in healthcare.

(2011:104)

In practice, guaranteeing these affects is a part of formal hospital training, which lays emphasis on the finer details of interactions and sensory components. Solomon, for example, describes a conversation he had with a staff member:

Gautham, who worked in the international patient services unit, explained that the hospital sponsors training sessions for its staff in 'cultural awareness' as it relates to national preference, and shared mnemonics he learned such as 'Americans need personal space' and 'the British enjoy silence'.

(2011:109–110)

It is worth mentioning that, over the years, much research in health geography has considered design issues and the reinvention, commercialization and corporatization of health-care settings (e.g. Kearns and Barnett, 1999, 2000;

Kearns et al., 2015). Although not focused non-representational aspects (on the immediate, moving or affective) but rather on representations (on meaning, metaphor and text), reading these studies and looking at their accompanying visual images, easily leads the reader to the opinion that the former are nevertheless important facets.

With regard to health research, as noted earlier, Kraftl and Horton (2007) reflect on a small conference where the findings of a study of young peoples' views on their health needs were shared with the participants who had provided the original data. The authors describe how both the event, and pre-planning meetings for it, were energetic and highly politically charged, taking on an affective life of their own, enrolling participants further in terms of their commitment and energy. The authors describe, for example, enthusiastic banter and practices that arose during pre-planning meetings:

F Why are you ironing?
P We're making some t-shirts Jo Yeah, we're just going to rip off the logo
F And you've been allocated the task because you're particularly proficient?
S It's his iron
F Oh I see. I like a man with an iron (2007:1017)

Affective health geographies: coping with chronic conditions

Although affect is recognized to induce feelings of wellbeing in 'healthy' individuals and groups in their everyday moments and lives (see Andrews et al., 2014; Coleman and Kearns, 2015), health geographers and others have also considered the roles of affect in living and coping with, and recovering from, chronic and long-term health conditions; mental health being an area of particular interest (see McGrath and Reavey, 2016). Duff (2016a), for example, explores 'affective atmospheres of recovery', showing the importance of three forms in particular; the first of these being sociability and social engagement. Duff's field notes reflect on a visit he made to one of his respondent's favourite urban hangouts, and the importance of the atmosphere there:

We walk into the café and Robert is greeted by name by two of the staff. Seated at a nearby table, a man about Robert's age looks up and nods. Robert says 'hi' and continues to a table adjacent to the windows overlooking the street. He sits down, stretches out and arranges his backpack in a way that suggests that this is his usual seat. I feel Robert relax. I relax too as I rest in the chair and take in the view. Robert starts telling me how he came to 'discover' this place and as he is talking one of the waiters comes over and tells Robert that another regular 'Paul' who will evidently be joining Robert later for a game of chess is going to be late for their game. Watching Robert speak with the waiter I am struck by how much more fluent his interactions are compared with our own stilted small talk as we walked the short distance from the tram stop where we had met to the café. I guess I am becoming more aware of the atmosphere Robert described in our first interview.

(2016:69)

The second form of affective atmosphere highlighted by Duff is safety and belonging. One of his respondents, for example, talks about her kitchen:

It's become my sanctuary, a place just for me, for cooking, for trying out new things but also I guess just for the way it makes me feel. Like, pulling into the driveway, it's like entering another world for me. I know that when I get into the kitchen and make a cup of tea it will be me here, that's me, and then the outside world. It (the kitchen) just makes me feel secure, in control of things.

(2016:70)

The third form of affective atmosphere is hope and belief. Duff recalls a particular walking interview he conducted with a respondent around an urban botanical garden where they together encountered and discussed a number of social, affective and material boundaries:

Nestled on the south bank of the Yarra River, overlooking the CBD and flanked by two of the city's busiest arterial roads, the gardens are a typically Victorian refuge from the activity of city life in Melbourne. As I walked with Liz from Flinders Street station over the bridge and into the gardens, I experienced a distinctive affective transition as I began to encounter a diverse range of human and nonhuman bodies. It is interesting to note how these encounters provoked the memory of previous visits and the affective states they engendered (like affective priming). Liz remarked, "I guess I see the Gardens as my own tonic, like on days when things get me down I know that if I visit the Gardens, and just sit for a while, practise my breathing, take in the light, I will feel better.

(2016:70–71)

Common to the types and uses of affect that Duff notes is the act of moving into or between different affective milieus (bodies being attracted to them and actively seeking them out). Indeed, Duff sees affect as a particular formation arising within an assemblage which gives the assemblage the ability to influence in a way which exceeds its basic human and non-human components. Their allure then is an example of them becoming more than the sum of their parts.

Whilst positive affection is important, other research has however noted how chronic and long-term illnesses have the potential to mediate affective engagements in unwanted and undesirable ways. This is undoubtedly the case for mental illness, but particularly notable here is Bissell's pathbreaking work on chronic pain. Bissell (2009; 2010) argues, for example, that pain has a certitude; being all-consuming, it blocks or diminishes the potential of surrounding affective intensities from creating positive affection in the pained body (and generally makes it difficult for the person to get excited about, and involved in, events taking place around them). Ultimately, as Bissell argues, pain can even lead to the pained body passing on negative affection to those bodies around it. Along these lines, a respondent to Andrews' (2018) study of living with Chiari Malformation (a brain disorder and mal-positioning that causes dizziness, disorientation and severe headaches) comments on their home life:

> Home is my anchor but also I can feel removed from what's going on around me. My kids playing, running around but me feeling out of it, in my own world, like watching it through a window.
>
> (2018, in press)

Both Bissell and Andrews consider research on affect to be important due to the possibility of it informing the design of interventions and strategies that might help open up the pained body to more desirable experiences (through participation in positive affective intensities). More generally, all three aforementioned authors – Duff, Bissell and Andrews – advocate for solutions to chronic illnesses that move beyond traditional approaches (that typically consider the physiology and/or psychology of the single chronically ill body), and instead consider the full assemblages involved in the making of health.

A final mention should also go here to the role of affect in how healthy bodies and healthy behaviours are sometimes constrained (as if they were unhealthy or problematic). Boyer (2012) for example talks about how the city affectively limits breastfeeding. On one level through breastfeeding women being made to 'feel' out of place in public space, excluded from the prevailing affective atmosphere. On another level through specific ideas and facilities, such as lactation rooms, which totally close women off from proximate affects.

Affective health geographies: health-harming and joyful behaviours

Health geographers have considered the affective qualities of environments and behaviours that are simultaneously health harming and enjoyable. Duff (2009) for example explores enabling spaces of drug taking in the city and their affective dimensions; 'enabling' in this particular context meaning both facilitating cultures of use but also of support, happiness and hope. A respondent states:

> Melbourne's such a buzzy city right! Like there's so much happening here, so much stuff to explore, it's so exciting. That's what I remember about the first time I visited here after school and that's really why I moved here. I just felt like I wanted to be in the middle of it all.
>
> (2009:207)

Drunkenness has also been a subject of attention, Duff and Moore (2015) for example arguing that the night-time city possesses diverse affective atmospheres that prime individuals to act in certain ways, thus impacting on alcohol-related problems such as injury. Similarly, Jayne et al. (2010) talk about public, media and policy concern for drinking culture which promotes violence and disorder in public space. In seeking a more nuanced understanding than mainstream research as to why people drink excessively, what exactly they do, what they get out of it, and how it is part of their everyday lives, she focuses on the infectiousness of nights out; the sense of togetherness and new/immediate social relations certain places offer (including brief encounters). A participant – a male between 18 and 24 – states:

> There's just nothing like that feeling you get on a Friday night ... it's like a freedom that you don't get at other times. You might go out to the cinema, or for a meal during the week, even for a beer or two, but on Friday and Saturday it's different ... I don't get much time to see my mates, I work hard and I'm often tired but on a Friday you get a buzz ... You look good, you've got your best gear on, there's nothing like that feeling that you've the whole weekend ahead of you to do what you want and the time to do it, and recover from the hangover ... and that everyone about your age feels the same ... you don't have to get up to work so you can go out and get pissed ... drink what you want, do what you want, go where you want ... and everyone on the streets is the same ... when else do you get that freedom to walk ... mess around, see and be seen ... be part of thousands of other people who are just having a good time ... you feel part of it, it's yours and you feel the places you go are aimed at you and the way you want to live. After a hard week at work it's exactly what I need.
>
> (2010:547–548)

This is of course, an experience and way of life that pubs, nightclubs and even urban planners facilitate through their purposeful provision of environments that sell an overall affective city experience.

On another level, research also shows how drugs themselves have affective qualities, altering users' receptiveness to certain affective environments and helping to enroll them in them. Indeed, Duff (2014a) considers how drug users consume drugs for their affective potential, two participants in his study commenting:

> [W]hen you're high like that it feels like your body is so connected, like every part is working perfectly, all coordinated and free. It's such an amazing feeling I love that part the most.
>
> (2014a:140)

It's so hard to put into words. But the way your body feels, these waves and rushes, especially if someone touches you, like runs their hands through your hair or something. I love it when my boyfriend does that – your whole body is so sensitive and every sensation is just this intense pleasure.

(2014a:140)

Certainly research does not elevate either place or intoxicant in thinking about how affects arise in these contexts, studies recognizing the mutually reinforcing co-existence of both. This, and further complexity, is articulated well by Tan (2010), who considers the role of affect in smoking and a range of infectious atmospheres smoking creates, adds to or detracts from (the human experience, they suggest, being relational, depending on one's attitudes and behaviours and preferences with regard to smoking). Indeed, Tan argues that for smokers, the act of smoking is itself affective, creating a new 'head space', but that for non-smokers smoking is an uncomfortable sensory experience which detracts from their preferred smoke-free affects. Tan argues that dedicated smoking spaces can be marginalizing and segregating to smokers who are banished to the peripheries (of workplaces, public spaces, entertainment spaces etc.). However, the author notes how dedicated smoking spaces can also be infectious environments of consumption, occupied with other smokers sharing 'exclusive' sensations and atmospheres.

One must also remember, however, that enjoying nights out and taking intoxicants, do not necessarily have to be detrimental to health and be categorized as problem behaviour, and can be purely joyful affective experiences. In Andrews et al.'s (2014) study, Sam – one of the researchers and co-authors – reflects on the highly commercial yet infectious environment of a cruise ship party she participated in whilst on vacation. The phrase 'all things in moderation' seems fitting here. The affective results of moderate drinking, coupled with affective qualities of the place and activity, helped Sam overcome her natural shyness:

While aboard a cruise ship I walk into a 'White Party'. The pool deck is filled with people laughing and dancing. Lights flash around and paint the deck in a rainbow of colours. The music is so loud that the bass booms in my chest, creating a new tempo for my heartbeat. **Boom! Boom! Boom! Boom!** I'm a quiet, shy person but the atmosphere puts pay to that. Everybody in white means everybody is in. I feel invigorated. My shyness melts in the hot summer air. I rush the dance floor to be with the happy faces. I dance, sing and move with them, my body moving in time, without effort and without thought. Minutes and hours pass unnoticed as I dance without fatigue. I carry on and on, and on. ... ...

(2014:218)

Affective health geographies: the politics and wellness in movement activities

Health geographers have focused on the affects involved in movement activities related to health politics, and also in movement activities that encourage both physical fitness and a mental sense of wellbeing. With regard primarily to the former, in an aforementioned study, Thorpe and Rinehart (2010), consider how social-justice causes (such as for environment or anti-violence) can be supported by alternative sports; not only through participants' developing coherent ideas, arguments and visions, but through their enacting the affective dimensions of the activities themselves. Indeed, the authors argue that participants in alternative sports frequently stress the embodied, sensory and immediate aspects of what they do; for them, merely talking about issues does not 'cut it' (see also Pavlidis and Fullagar, 2013, on roller derbys, and Garfield, 2012, on parkour/urban free running).

With regard primarily to the latter, Macpherson (2008), for example, considers the role of laughter and humour in walking groups comprised of visually impaired people (as the author suggests, undertaken to assist their cardiovascular health, to evoke an immediate sense of wellbeing, to build interpersonal relationships, to reduce nervousness of unfamiliar environments, and to liberate individuals in the face of stereotypes and bodily challenges). Macpherson argues that, during walks, humour – using intellect and often narrative – arises consciously and reflectively, whilst laughter – often the response – arises less-than-fully consciously as an embodied and contagious phenomenon connected also to the embodied muscular practice of walking. Specifically she comments on her participant observations:

> Belly laughs, titters, giggles, anxious exhalations, and guffaws would reverberate through the country air on days out walking, creating a transient sonic element of the landscapes we passed through. This laughter would occur in response to embarrassment, the slapstick, the incongruous, and the paradoxical, the subversion of stereotypes, trips and slips, at self-mocking tales, in-jokes and laughable laughter... Ellen [says to the group as they start off again] 'Right you blind buggers' to which many other walking group members laugh.
>
> (2008:1081)

In other research, Simpson (2017) considers the affective atmospheres important in commuter cycling as a form of active transportation. In his study, cycling infrastructure is considered to be more than an inert context, and to have its own agency and affective capacity. Simpson highlights Merriman's ideas on this; that something as banal as a car park, for example, is not bland, static and neutral, but can be a complex environment which possesses multiple affects and other qualities, fed by, and feeding, bodies and objects (Merriman, 2016). Meanwhile, Barnfield (2016) talks about the infectiousness of recreational running; running beside people, in front of people, behind people, all being part of a collective experience. His field notes state:

> As I arrive I start to see all the other people. Hundreds and hundreds. There are tents set up for all sorts of things, there is music being played, and there are gift bags given out. Lining up to start the run I am in the middle of a sea of people who are jumping up and down, stretching, and chatting nervously to friends. The countdown is on and we are off. Two loops of the 5 km track through the more floral part and the wooded part of the park. People are jockeying for position as we start to run. There is lots of noise as the crowd is large. The pace is fast but I focus on my movement. Hitting the stride and staying with a group before moving on up the path. In the wooded part of the course it feels eerie. Then we burst out into the evening sunshine and hear the clapping of the people. I am feeling tired after one lap but the energy from the spectators keeps me going. I find myself pushing for the finish line to the sound of my number being read out. I drive forward, skipping across the tarmac. I stop, receive a medal, and am left to find my belongings and absorb the noise, the colour, and the positive sensation of moving along with bodies of different sizes, shapes, and speeds.
>
> (2016:286–287)

Affective movement activity is a popular area of inquiry, other examples studied by geographers, that evoke wellness and wellbeing, including swimming (Foley, 2015), kayaking (Waitt and Cook, 2007), surfing and windsurfing (Evers, 2009; Humberstone, 2011), whilst related research has also begun to consider bodily functions and attributes that mediate affective transfers between bodies, such as sweat (Waitt and Stanes, 2015).

In sum, affect is increasingly being explored and deployed as a key concept in more-than-representational health geography. Beyond the particular fields of scholarship described in this chapter, there remain many other areas of health and health care where it might help explain and convey important collective, energetic, and transpersonal aspects.

5   Characteristic styles and priorities of more-than-representational health geographies

My son Justin asked me what I was working on. After I struggled through an explanation of non-representational theory for eight year olds, having no formal social science training, but lots of lively engagement with the world – not to mention an open mind – he immediately rolled out a bunch of active relational qualities and contrasts – or 'things' in his own words – that I could look at, excitedly adding to his list 'hot/cold, up/down, fast/slow, smell/no smell, light/dark, shiny/dull, heavy/light, round/pointy, straight/curved, here/there, rough/smooth, coming/going, open/closed'. Finally, and remarkably, he then added 'not love dad, that's too complicated!'

The observation above was typed during the early stages of this book. At the time I was taken back at seemingly how easily a then eight year old could get at least some of what NRT is concerned with, particularly when so many academics – not to mention myself initially – find it to be challenging. Building on Justin's observations, this chapter outlines five key concerns that together create the overarching style of NRT; practice and performance, foregrounds and backgrounds, senses and sensations, impulses and habits, and ordinary and everyday. These are not an aesthetic of research but, reaching to the very core of studies, they entail the fundamental things that are looked at and the way the world is apprehended and engaged (for other typologies see Thrift, 2000, 2008; Cadman, 2009; Vannini, 2009, 2015b; Andrews, 2014a; Andrews and Grenier, 2015). Indeed, these five concerns might be thought of as substantive features of the world; qualities that characterize the immediacy of the human condition and experience, that NRT researchers seek to animate in their empirical studies. After explaining the basis of each, the chapter elaborates how each has arisen in explanations of health in more-than-representational health geographies.

Practices and performances

NRT perceives a world that is industrious and productive to greater or lesser degrees at every single moment. Hence a concern of the approach is to focus on the everyday practices and performances that make it so. As Vannini (2015b) suggests, this diverges from a primary interest across many traditional forms of social science research with thought (i.e. discourse, ideology and language, evident in practical concerns for outlooks, states of mind, views, ideas, values, beliefs and so on). Practice and performance are instead about singular and collective efforts exerted in particular ways to achieve particular ends; the expressive arrangement and engagement of bodies and objects as 'lines of action' with different intensities and directions. Moreover, NRT considers very basic and active elements of practices and performances – including intervals and spacings between bodies, objects and the events they create – whether they be habitual or irregular, intentional or coincidental, scripted and expected or unscripted and unexpected (Vannini, 2015d).

On one level, researchers consider signs and meanings in practices and performances that might be consciously read by humans – including themselves – and that, in turn, reach full signification. However, as Anderson (2009) suggests, they also realize that practices and performances do not necessarily have to move into the realm of full signification to be effective and powerful. That, through humans participating in practices and performances, they simultaneously communicate to and learn the world less-than-fully consciously; once becoming stable, the resulting visual geometries helping make the world familiar and comprehensible to them (Thrift 2008). In sum, as Vannini (2015d) notes, using the analogy of a theatrical play, NRT is not primarily interested in what might be going on behind masks, costumes and scenes. It is more interested in actual masks, costumes and scenes themselves.

Practicing and performing health geographies

Certain research in health geography is focused on the practices and performances of complementary and alternative medicine (CAM), and specifically those that constitute the daily work of practitioners of various modalities. Although varied and often unique, these are regular and familiar enough to define each modality and provide a certain consistency of experience. Hence, these are practices and performances that are often carefully thought out, justified and in some cases regulated. Patterson (2005), for example, describes how reiki practitioners put their hands on and off their clients' bodies at specific times, and the distinctive hand movements involved in the affective transfer of energy between them and their clients. Elsewhere, in Andrews et al.'s (2013) aforementioned study of CAM more generally, therapists stressed the importance of body poisonings and body relationalities even before the formal treatments begin; the way in which these might be less-than-fully consciously read, and their consequences for the remainder of the therapeutic encounter. A respondent in this study explains:

> I think you have to create the space and time, and you have to communicate that with your body language. You have to sit down. I mean if you're

Figure 5.1 Practice and performance

standing and you've got the chart under your arm, they know they've got three minutes and you're not going to go where you want to go. I think letting the person feel that they have the time is important. If you can convey that through your words and tone of voice and your body, and try to create the time space, that is critical. You could probably do it in the broom cupboard if you had to.

(2013:106)

Other CAM research is focused less on practitioners and more on consumer practices and performances. Notably, McCormack (2013) animates, through words, written diagrams and photographs, the lines of movement in dance movement therapy. Participants experimentally freeing themselves from their self-consciousness and, through bodily gestures, less-than-fully consciously forming new shapes that are their own unique expressions of the prevailing rhythm. Indeed, as McCormack describes, these are practices and performances that incorporate increases and decreases in energy, create their own territories in space, and make dialogues in space, all of which are often less-than-fully consciously half read by others. In terms of therapeutic qualities, they are forms of creative expression with momentarily liberating, spiritual, and mentally offloading affects for those involved.

Certain research in health geography looks beyond practices and performances in formal therapeutic settings – such as the CAM clinics and dance studios noted above – to those conducted in public space. An important theme here is the negotiation of everyday landscapes through movement activities themselves often motivated by potential fitness or therapeutic outcomes. Middleton's (2010) study of urban walking shows, for example, how walkers practice certain techniques to maintain their rhythm and momentum through the city, including timing their approaches to and across walkways, pushing forward across junctions, dodging other people and cutting corners – sometimes consciously, but often on 'autopilot'. Taking these ideas further, Lorimer (2012) describes how long-distance running involves these kinds of tactics which, being more than physical expressions, involve mental practices and performances set within an overall state of absorption. Specifically, he talks about this in the context of negotiating hills, and in terms of the 'kinetic intelligence' drawn on when negotiating snow-covered terrain:

> Slopes are, of course, in simplest terms, kinds of surface set at an angle from the horizontal, but they do generate a different intimacy, and inter-relation, between runner and world. The way up a long hill needs to be searched out, steadily, most surely. On occasion it amounts only to a painful protest march. But I know I've found the quality of movement when everything tunnels, and patterns of thought are trimmed back enough to become single-minded. The gradient helps this happen, of course, as though disinclined to let attentions wander. My field of vision shortens to the next few steps only; looking far ahead of oneself can prove overwhelming. Gradually, the experience of ascending is encapsulated. That is, sensations close down to small matters, only of most immediate concern: foot placement, control over breathing, stride length, rhythm and pace.
>
> (2012:84)

At first under-prepared, legs buckle up and bow, spring and step fall out of sync. It's knockabout stuff, somewhere between carousel and big dipper. When body position is thrown, shape must shift, so that the balance is re-jigged, just to allow the next step to fall. There is a biomechanical explanation for such a loose-limbed assembly of kinetic intelligence. Uncertainties in surface and irregularities in its resistance set off sensory receptors – science calls them proprioceptors – of which a runner's

deliberating thoughts know precious little at the time, and which transmit information first through nerve endings within muscles, tendons and
joints. Acts of recovery and pre-emptive repositioning trigger instinc-
tively, flickering between eyesight, cerebellum and moving body. Seen at a
distance by solitary dog-walkers and families out sledging, my stuttering
progress must have looked comical and curious.

(2012:86)

Notable here is Lorimer's point about display, their being potential observ-
ers to such practices and performances, the judgements they might make and
conclusions they might reach. Indeed, this theme is also important in other
research focused specifically on how the non-purposive practices and performances of 'different bodies' play a part in how they are othered and marginalized. Colls (2007) for example explores the materiality of fat bodies, paying
attention to their surfaces and gestures; their 'swaying, hanging, lifting, rising
and dancing' (2007:361). These are performances that are focused on by those
who judge fat bodies – including media and policy makers – who attempt to
limit and contain them in various ways. Colls notes that fat bodies are simply
doing what fat bodies do but, after this, they are judged by others to be speaking to them and society in particular ways; opposite ways than, for example,
they might consider walking and running bodies to be speaking to them.

Foregrounds and backgrounds

NRT is concerned with the overall physical and sensory picture; both the foregrounds and the backgrounds in life. Foregrounds are the happenings and
unfoldings that dominate and make a moment. As Vannini (2015b) lists, they
are the physicality and conscious registering of:

accidents, predicaments, advents, transactions, adventures, appearances,
turns, calamities, proceedings, celebrations, mishaps, phenomena, cere-
monies, coincidences, crises, emergencies, episodes, junctures, milestones,
becomings, miracles, occasions, chances, triumphs, and many more.

(2015b:16)

These reveal in different ways, often signifying the emergence of something new, and moments of new potential (Vannini, 2015b). Moreover, the
new courses they reveal, whether indeterminate or determinable, are often
irretrievable and often bring forth conflict, drama, certainty or uncertainty
(Dewsbury, 2000; Vannini, 2015b).

Backgrounds meanwhile are the many things and events that fall outside
our full awareness, being less-than-fully consciously registered and/or taken
for granted in environments, such as lighting, shadows, colours, temperatures,
or body and object events either too far off to discern or not in immediate
focus (Vannini 2015b). They frame events, yet their background status belies

Figure 5.2 Foregrounds and backgrounds

the fact that they possess subtle power (as in their contribution to affect), helping make the mood and determining the energy of places. Backgrounds might be stable, or appear to be stable, yet importantly they are often mobile and almost always open to some form of manipulation. Relationality and positioning is also important. Depending on focus, one person's foreground can be another's background and, from any particular vantage point, foregrounds and backgrounds might bleed into each other or, over time, one might transform into the other.

Foregrounds and backgrounds in health geographies

Three key areas of research in health geography showcase the importance of foregrounds and backgrounds in health scenarios and contexts. The first of these is the purposeful design and manipulation of health-care settings.

With regard to CAM, for example, in Andrews' (2004) study, therapists comment on the specific designs and decorations they introduced into their clinics – including visual and audio interventions – to help create a soothing and relaxing atmosphere and help their clients drift away. Elsewhere, focusing on the conventional sector, Evans et al. (2009) consider the therapeutic potential of landscape art in hospital waiting rooms. These authors draw on Lacanian theory to explain their empirical observations. On one level Lacan's idea of 'the real', which challenges the notion that the physical surface of life is superficial, and instead promotes the idea that it exerts forces on humans that are alluring yet are in themselves meaningless (i.e. cannot be attributed primarily to language and knowledge). On another level, his idea of 'the gaze', which escapes the idea of the human view being all-powerful and objectifying by bringing the object itself into play, which can incite its own positive or negative reactions by staring back at humans. Indeed, in Evans' et al.'s study, the medical environment is a real and slightly disturbing background (a cold and austere space) countered by the comforting, calming and pleasing gaze of art (a background that moves into foreground once noticed). Summarizing, Evans et al. quote Lacan's ideas about how comfort and pleasure is found in the aforementioned processes:

> [S]ociety takes some comfort from the mirages that moralists, artists, artisans, designers of dresses and hats, and the creators of imaginary forms in general supply it with.
>
> (Lacan, 1992:99)

The second of these research areas is mental foregrounds and backgrounds in health-care practices. Like Evans et al., Andrews and Shaw (2010) used Lacanian ideas and theory to articulate the foreground of needles in needle phobia: their negative gaze (how they stare back at needle phobics) and the diversionary visualization practices nurses employ tactically in the moment to counter this:

> In the moment you have nothing else open to you, no other options other than to visualize. Patients will feel this themselves and, if you don't react, they will grasp saying 'talk to me, talk to me about something'. And my response could be a joke or even about somewhere. Even if it is not that extreme you can see that they want to be somewhere else. One patient turned to me and places in my life. She said in a panic 'tell me about your neighborhood in the council strike. What did it look like?'
>
> (2010:1807)

In Andrews and Shaw's study, the disturbing 'real' foreground of needles falls into the background, failing to gaze, once diversionary therapeutic foregrounds are introduced by nurses. Meanwhile, in Andrews et al.'s (2013) aforementioned study of CAM, a psychotherapist talked about setting the scene in her visualization practices. Mimicking one of her therapy sessions, she narrated how she paints the scene for her stressed client, including some finer details and incidents, with words:

> You are walking down that road again, the sun is out, but this time you take the other route, the longer quieter way to work. You calm yourself down as you take in the nature, walking more slowly, seeing the trees, the flowers, the birds. You feel more confident this time, reflecting on your approach, how you are going to be. You arrive at work, the same office, desk, people busy… things around you, but you feel different there. You can see things for what they are, are able to make the changes, able to deal with people'. You see, you move them through, keep them onboard, keep them going with you! No time for over-thinking. I have to go along, go with the flow, to feel the journey so that they will.
>
> (2013:106)

The third and final area of research is broadly focused on backgrounds in community-based spaces for health and wellbeing. These can equally contribute to, or result in, positive or negative experiences. With regard to positive, Conradson's (2007) study of therapeutic retreats, for example, stresses that in order to obtain a feeling of 'stillness' (important as a mode of feeling and the basis for new clear conscious reflections) complete auditory silence is not necessary in these particular settings. Rather, Conradson suggests that low-level background noise can greatly assist. With regard to negative, Lorimer (2012) describes a particular running event and the rather disorientating experience of the regular landscape lying ahead of him:

> It's not only changeability, or difference, in surface that can register as remarkable. Uniformity and regularity can have startling effect too. A few years ago, while holidaying on an island in the Baltic Sea, I found a run of twelve kilometers that was utterly unwavering. At least, that's how they appeared on the map. Not a curve, or a bend, nor any up or any down: just the straightest of lines, made by a minor road. And it looked that way when I arrived at the spot, early one morning. The run, in its entirety, seemed visible, defined by principles of perspective: a vanishing point, the rising sun, the painted dashes marking the mid-point of the metalled surface and the tall conifer trees that bordered its verges. Some while after setting off, a queer kind of dizziness came over me. The squared-off scene blurred at the edges, and I lost resolution too. If my legs were running, then my head was swimming.
>
> (2012:83–84)

As evidenced in these studies, this particular area of research has also shown the importance of different sensory attributes in feeling foregrounds and backgrounds (in the above cases sound and vision in particular), either directly or as an affective resonance of possible happenings. This leads us to the following section.

Senses and sensations

Senses and sensations are important to NRT on two levels. On one level because NRT recognizes that knowledge is not only something consciously learned, but also something formed processually through the body. As Roe and Greenhough (2014) suggest, this position somewhat echoes that of Bourdieu's habitus although, whilst Bourdieu stresses how human embodied perceptions of, and reactions to, the world are shaped primarily by external social positions and structures (that have been internalized by the body), NRT instead lays greater emphasis on how embodied perceptions of, and reactions to, the world are shaped primarily and 'purely' by the internal/natural/biological sensual properties of the body, and specifically through the body doing things within assemblages of other bodies and objects (more akin then to the often used sociological ideas of embodied learning and embodied knowledge; Crossley, 1995, 2004). Meanwhile senses and sensations are important on another level because, as described earlier, NRT takes a distinctly relational materialist position, and hence the key parts that human bodies play in greater material events and processes are thought to be critical: in other words, the parts body registers and felt sensations (singular or shared) play in the making of life's practices and performances, foregrounds and backgrounds, impulses and habits, ordinary and everyday and so on.

In sum, then, NRT's emphasis on senses and sensations is part of an academic engagement with the body, not so much concerned with how the body is constructed (as in biology and other health sciences), but instead concerned with the 'being' and potential of the body through material relations (Cadman, 2009). Indeed, this is an engagement which realizes the body as a mediator in the non-representational world. Senses – auditory, gustatory, olfactory, visual and touch – picking up and registering the world and driving human responses, they being critical to how the body co-evolves with things (Merleau-Ponty, 1962; Vannini, 2015b).

Sensory health geographies: ways of knowing medicine(s) and health care

Three main bodies of work constitute health geography's engagement with senses and sensations, the first of these being focused on experiencing and knowing types of medicine and health care. In Andrews' (2004) aforementioned study of practices in CAM, for example, therapists emphasized the importance of locating the body through the senses in its immediate environment, a psychotherapist who was interviewed commenting:

> One place that psychotherapy helps is in accessing your own body. I allow people to earth themselves in their surroundings, for example, reminding

Figure 5.3 Senses and sensations

them to be aware of their chair, and its contact with their body. When dealing with a very active mind, it can be calming and stabilizing to remind it of its physical sense and its familiar sensations.

(2004:312)

Moreover, Andrews et al.'s (2013) study of CAM found that therapists laid great emphasis on sensory aspects of their immediate practice environment. Specifically, one comments on how warmth was achieved through the use of specific objects and their textures:

One of the things that is most important to me is that the linens are natural fibres. That's like up there because, in the work that I do [manual bodywork], one of the things that is very important for someone's own healing capacities to be generated is warmth. When you use synthetic fibres there's a cold quality to them. And for what I want to do in my treatments it makes a huge difference for someone to relax when they're warm, rather than use that extra energy to warm up. And also my oils. My oils tend to that warmth element as well. They bring a quality of warmth to the skin.

(2013:105)

(see also research on the role of smell in engagements with the CAM practice of 'care farming'; Gorman, 2017).

Certain research has focused on conventional medicine and health care. Considering the scientific work that goes into informing bio-medicine and justifying its methods, Greenhough and Roe (2011) explain, for example, how in animal laboratory work, clinical trial nurses, vets and animal caretakers have in common the ability to sense and respond to both verbal and non-verbal signs of distress and suffering in the animals they care for. The authors suggest that these individuals look at, listen to and touch animals but oftentimes these are less-than-fully consciously occurring 'sensings'. The authors argue that these sensory aspects are a critical part of becoming articulate in care, and part of an ethics of care. Indeed, these are known to be aspects of practice that are equally evident in the affective relationships between nurses and human patients in traditional hospital settings and training environments (Ducey, 2007, 2010; Soffer, 2015). Elsewhere, Solomon's (2011) aforementioned study of affective relations in Indian medical tourism also stresses the importance of sensory experiences. Particular sensings of medical environments were highlighted as important so that patients – often visiting them for the first time from far afield – recognized them as safe, familiar professional places. However, the authors also argue that patients' negative sensory experiences of the broader urban environment equally come into play and have to be accounted for by hospital staff. In this regard, the researcher notes:

Judy thought the staff in the hospital were kind, and that they take better care of a patient than in the US. But these advantages, she said, had to be put alongside the challenges of their first visit to India, including their responses to crowds, poverty, and a different set of sense stimuli: 'Some of the smells are really horrific, and it's hard to get past the smell to eat...I don't mind trying different things... it's just Indian food ...the smell.' These criticisms were each prefaced by an apology to Gautham, who stood by Bill's bed and responded with a polite nod of acknowledgment. Gautham assured them that these issues were commonplace for Western visitors and that Icon prides itself in being able to offer specially catered meals.

(2011:110)

Exploring and exposing experiences of health care has been an important objective of humanistic health geography; research here being focused on personal, group and political processes of meaning making and identity formation (Kearns and Barnett, 1997; Parr, 2003). What this current more-than-representational research adds to it is an appreciation of the less-than-fully conscious bodily perceptions involved in parallel processes of knowing places of health care on different but equally powerful levels; how felt 'sensings of place' add to known 'senses of place'.

Sensory health geographies: thinking and moving in outdoor spaces

The aforementioned example of Indian medical tourism nicely previews the second body of work in sensory health geographies which is focused on movements and negotiations in outdoor/public spaces. Lorimer (2012), for example, describes his multiple sensory engagements with changing landscape in his hobby of running. Here, the sensory becomes familiar to him and, although not often consciously dwelled upon – particularly after the fact – is nevertheless an essential part of his performance, the overall event, and his experience of it:

> As much as there might be stretches of country that can induce in me a powerful yen or sense of belonging, there are too, unnumbered surfaces, forms, textures and gradients to which I have become deeply attached and with which on revisiting I find a real rapport. Less scenic, more sensed... ... Leafing through this atlas of remembered surfaces, I might pick out a medley of terrains. There is the spine-tingle and earth-shatter of a scree slide, the winter drag of stubble fields, the odd experiment barefoot in wet grass or as a long shore drift, the past times of peat hags making legs feel spring-loaded or the high-steps demanded by heathery hillsides, the magical box-fresh bounce encased in new footwear, the cyclical hypnotism induced by the athletics track and the unforgiving rubberized rip of the gym treadmill. Some surfaces are harder to put a name to, or to easily place; still impacted in memory but inchoate and thinned out.
>
> (2012:83)

Similarly, Brown (2017) employs video ethnography to consider the role of 'ground feel' in walking and cycling; the way that sensing textured terrain provides emotional and physical experiences which are important in motivating and sustaining regular exercise. Specifically, Brown talks about textual immersion:

> Most simply, how the ground feels through the body was central in generating a range of pleasurable somatic sensations, such as feelings of dynamic pressure, massage, vertigo, vibration, measured disorientation, equilibrium, 'flying' and 'springiness' and feeling held. Furthermore, it was clear that this ground-feel of dynamic touch generated in many

participants a sense of proximity and immersion and a direct contact or intimate interactive connection with their environment.

(2017:311)

Thinking of the practical lessons and consequences of his research, Brown concludes that these types of experiences need to be taken seriously by those designing and/or providing opportunities for outdoor exercise. Specifically, the author argues they need to resist thinking about 'good' or 'bad' textures and terrains (as this is too simplistic), and avoid producing over-controlling and over-standardizing environments.

Other research in this field showcases the role of senses in the interplay of thinking and moving. Macpherson's (2009b) study of walking groups for the visually impaired considers, for example, the role of touch. She suggests that in this particular group with their particular disability, their touch replaces their sight in multiple ways: not only in knowing what immediate situation one is in and how to move forward successfully, but also in mentally/spiritually immersing oneself in the landscape, and the visualization and contemplation that involves. Two participants in her study comment:

> Well texture, texture is very important....and obviously you can feel through your feet the texture of what you are walking through... and because you can't visualise where you are about to put your foot that's important, and like I said yesterday ...it is like your knees act as shock absorbers, so your body sort of takes on the role almost of an extra hand, you can feel through the shoe, you are feeling the texture, instead of anticipating with the eyes and generating images with the brain... you have to analyse the texture and feel with your body.

(2009b:179)

> [I] gather my own impression of what the terrain is like – I get that through my own feet more than anything else. You know?... actually you can pick up a lot of information from your feet and maybe because I used to see it helps piece together a picture of what the whole thing is like just from walking over it.

(2009b:180)

Similarly, diarists/participants in Middleton's (2010) study of urban walking noted the multisensory dimensions of the activity and their relationship to both physical progress and internal thought processes. One commented specifically:

> Monday morning the bin men throw the contents of the bins into the road for picking up later. This makes things very unpleasant to look at – half the bags seem split. This route has many quiet sections which erupt

into noise at the main roads, Kingsland Road, Hackney Road, Bethnal Green Road. Otherwise it's a good way to contemplate. Not many other walkers around. Some women walking dogs, a couple of old boys on benches (alone).

(2010:584)

Sensory health geographies: sensing in and of the body itself

As a few of the examples described previously allude to, certain sensory experiences move beyond the core five senses that are initial connections between the body and the environment (sight, hearing, smell, taste, touch). Being bodily sensations in and of the body itself, some of these are associated with basic body needs and functioning (including hunger, thirst, pain, spatial and temporal position), whilst others are how emotions (such as fear, joy) are physically felt (such as increased heart rate, adrenalin rush, sweating or trembling).

With regard to the first aforementioned internal sensory group, geographers have stressed the roles of, and relationships between, human and non-human actors. With respect to pain, Bissell (2010) talks about how bodily pain spans the conscious and less-than-fully conscious in life; how it pushes physical human limits, affects our engagements with the world and messes with our reasoning. Bissell suggests that we can talk, for example, about how pain physically feels, but how it feels to live with pain is a conscious consideration arrived at through our affective experiences. Elsewhere scholars address the pain in certain movement activities, which is often overlooked by studies that focus on movement's pleasures. Wylie (2005), for example, talks about the pain experienced in the effort and practice of long-distance coastal walking being more than internal; more 'in and of' the landscape itself:

> New postures and surfaces materialize and new affectual extensions resonate. The bone pain of walking is realized in an aching halo of landscape, with the ground immediately beneath your feet and the slope climbing above and the coast unspooling relentlessly ahead. Pain occurs neither 'in me' nor 'in that' – the externalized body – but 'between me and it', in this step, this next step. And so the landscape emerges as malignant.
>
> (2005:244)

Similarly Spinney (2006) recalls cycling up a French mountain:

> Time and tarmac pass. I hear the whir of my gears, the constant rhythm of my breathing, the contact of tyres and asphalt; I feel the dull ache in my legs, the emptying and filling of my lungs. There is nothing to do but fix your eyes on the road and think of something else.
>
> (2006:722)

With regard to the second aforementioned internal sensory group (physically felt emotion), Bøhling (2014), for example, considers the experiences of urban nightlife aligned with drug alcohol and use. A participant in this study talks about the joy in taking drugs in various parts of clubs, both the core sensing of the environment (such as hearing noise and seeing light) and the drug-induced bodily sensations being important to them:

> I think it's great when the music is very loud, because when it is lower people will start talking, and then you'll hear this 'bzzz' of voices which will disturb me in my dancing, so yeah, I think it's great when I can feel it in my body, but it has this positive side effect as well. It's also that [in the main room] the music is much harder and everything is darker so people do not have as much eye contact, and when you go to the other bar, where the music is softer and there is more light and there is this sexy vibe, people talk and flirt more, which is also nice I guess, but it's not for me.
>
> (2014:376)

Also conveying physically felt emotion are a number of studies exploring fear-related sensations related to phobic encounters (such as sweating, shortness of breath, dry mouth, shaking, pounding heart) including in phobias of needles (Andrews, 2011), heights (Andrews, 2007), social situations (Davidson 2000, 2003) and nature (Smith and Davidson, 2006). An acrophobic respondent in Andrews (2007) study comments, for example, on his experiences at height. He could visually see he was at height, yet, beyond that particular core sense, his sensations were wide-ranging:

> My body feels heavy and it's very hard to move. You have to move slowly and very carefully. My vision seems to narrow and I sweat.
>
> (2007:312)

Andrews argues that, with such a comprehensive engagement, the sufferer becomes part of the overall acrophobic event, their sensory responses to their physical situation feeding their physical and behavioural bodily reactions which are palpable and obvious.

Impulses and habits

> Habit is seen to push understandings of agency away from the 'existence of a sovereign wilfulness' towards a distributed understanding of the performance of subjectivities as more or less durable but ever-changing dispositions, potentials and failures.
>
> (Bissell, 2015:50)

Impulses and habits are subjects of particular interest and relatively recent additions to NRT literature (Dewsbury, 2011, 2012, 2015; Dewsbury and

Bissell, 2015; Bissell, 2011, 2012, 2013, 2015). In terms of basic definitions, whilst impulses are the biological and/or psychological bodily pushes or urges of varying intensities that might prompt action, habits are the actual actions themselves when repeated over durations (ranging from simple movements to complex events). In terms of what they do and how they work, either one can be experienced consciously or less-than-fully consciously. Habits might be based on impulses to act in accordance with particular meanings or to experience particular sensations, and/or they might be more subtly evoked by the physical or mental feel of, familiarity of, or learning of, repetition itself. Whilst these bases are important, so is the 'automatic' nature of habits, as less-than-fully reflective 'modes of being'. As Kraftl (2013) describes, being produced and felt between people, habits play an important part in collective action and experience; they helping create the base rhythms of humanity. For example, the collective affective experience of a particular practice might act as the force behind its repeatedly happening amongst people, and hence becoming collectively habitual (Kraftl, 2013).

In terms of how understanding impulses and habits in these ways potentially changes academic thinking, Dewsbury and Bissell (2015) and Dewsbury (2015) argue that habits challenge scholars to think about how (i) the very notion of a sovereign, autonomous and willing subject might be flawed, as people and their actions are shaped by wider forces (i.e. places and people emerge through passivity to affection and habitual performances), but conversely how (ii) habits might be regarded as a form of intelligence that allows humans to act without conscious effort in familiar situations; moreover, how (iii) habitual performances might be one process through which people gain sense, understanding and awareness of the world (and a sense of being through their parts in particular becomings), these senses not being 'theirs' but instead distributed within their milieus (one part, for example, of Bourdieu's notion of habitus, whereby engrained performed habits both lend us capital and help us navigate social worlds); how (iv) the past, present and future are linked through biological and affective forces that join them in habitual performances; and how (v) in the vitality of life, something new is created and subtracted from habitual performances; something other than and more than what is itself repeated. Indeed, as Tucker (2010a) observes, this is consistent with a Deleuzian view that emphasizes life being created through constant variation; habits here envisaged, not as people engaging in life with a particular fashion, but as an essential part of life in the making.

In sum, the idea is that through these lines of thinking we might move beyond the popular idea of habit as something specific (narrow) and problematic (such as an addictive behaviour), or the academic idea of habit as a negative, unintelligent force that obstructs change and positive potential (allowing, for example, institutional power to be enacted over malleable minds) (Lea et al., 2015). Moreover, these lines of thinking allow us to modify even current critical perspectives on habit. As Roe and Greenhough (2014) point out, Bourdieu's thinking on habitus, for example, attributes habit formation to underlying social structures (such as social class, status and other hierarchies). Instead, as Roe and Greenhough suggest, NRT challenges this understanding by laying greater emphasis on more-than-human processes and materialities, providing a more fundamental and more fluid notion of habits and the making of culture.

Impulsive and habitual health geographies

Perhaps unsurprisingly, traditional health geography includes a focus on habits, mostly in terms of addictive habits – such as smoking, drinking and taking drugs – that lead to poor health outcomes (most of this research focused quite broadly on lives lived by, and support for, those affected). More-than-representational health geographies, however, take this interest further, focusing either on the habitual daily routines of those affected and/or on their sensory experiences of the impulses involved (such as cravings or pain from damaged bodies). With regard to habitual daily routines, Evans (2012), for example, examines experiences tied to a managed alcohol program (a 'wet' supportive living facility) for recent homeless street drinkers. Evans describes how residents entered or remained in the program out of fear of consuming an even larger amount of alcohol if they did not; a virtual fear made real by their affective knowledge of that potential future. Thus they actively wanted to escape bad habits and the places that facilitate them, and instead move into new places with new regimes that facilitate good habits. Evans reflects:

> Troy who I interviewed at a downtown homeless shelter, endowed the prospect of moving into Mountainview with immense hope: 'I'm gonna start going straight. I'm gonna be clean all the time, I hope. Think it's a real good idea for me. I'm gonna get up to the mountain [facility], I'm not coming down here no more [downtown],' he declared.
>
> (2012:194)

Services and facilities can hence themselves become part of a new overall habitual life routine. Indeed, looking at this idea further Evans (2011) examines the political dimensions of a low-barrier (wet) emergency shelter. His respondents talked about the shelter as a place of support and recovery where they go for some downtime between binges, whilst a respondent – a staff member – reflected on how users repeatedly went 'in out, in out, in out' (2011:27) in a repetitive cycle as part of their lives.

With regard to impulses, Wilton et al. (2014) considers how abstinence-based treatment utilizes the design and regulation of the setting to re-build an individual's ability to exercise self-control. A director talked to the researchers about the dispositions of those entering his programme and how its routines produced better habits (replacing repeated violent and painful experiences, for example, with order, friendliness, cleanliness and self-care):

> You cannot think your way into action but you can act yourself into think-ing ... People especially in early treatment tend to be in a fog. They're unable to make decisions. They're unable to carry forward anything ... So we give them a very regimented set of rules to follow. We make them get up in the morning, we make them shave, we make them have their breakfast, so they just put one foot in front of the other and we do that for the whole 21 days and at some point they'll say 'hey, you know what? I'm feeling better'.
>
> (2014:296)

Other research moves well beyond addiction, scholars considering every-day public spaces and practices of mental and physical wellness. Middleton's (2011) aforementioned study of urban walking, for example, describes much of this practice as an embodied habit involving routine and ritual, both to get out of the house and participate in the first place (whether to walk, when to walk, what route to take) and a key part of journeys on foot (less-than-fully conscious decisions and actions made to maintain momentum, keep on walk-ing and not give up). Elsewhere, Lea et al.'s (2015) study of mindfulness medi-tation considers these holistic practices both as habits and as self-reflective moments/positions, that involve agency that is spread between the body, mind and context and a multiplicity of space-times. The authors argue, for example, that one might consciously employ reflexive mindfulness techniques as a posi-tive habit set within the flow of everyday routines in order to interrupt or keep at bay negative thoughts and emotions that themselves have often become habitual and might otherwise cascade into more serious mental health prob-lems. One of their participants states:

> I think it's actually noticing your thoughts and not letting them run away with you all the time, ... I think that's what has happened since I've had the breakdown is that I was one of these people whose thoughts raced a million miles an hour. They still do to a degree, but much less so. Now that could be the medication ... But maybe the combination of medica-tion and meditation slowed me down for the first time ever really.
>
> (2015:55)

Indeed the authors suggested that participants were trained to eventually notice when they had entered bad mental habits on 'auto pilot' and to generate the alternative habit of awareness of the present. The authors conclude that:

> Clearly [these] habits are not hidden, mysterious and beyond agentic intervention, but neither are they easily accessible and readily mutable.

Rather, agency can be seen to shift between habits – which, as they repeat, reproduce corporeal and cognitive regimes – and the reflexive self who manages successfully to deploy a particular technique enabling them to change their relation to these habits.

(2015:61)

Finally, there is a focus on habits common within private spaces. Tucker (2010a), for example, describes how individuals suffering from mental distress and illness activate repetitions in their home spaces. These become habitual creating a form of stability in their lives and provide a solid anchor to help them make sense of the otherwise constant mental fluidity of their worlds. As Tucker argues, their spatial and emplaced habits create the sameness of home for them, this enabling them to enact variations around predictable anchor points that might empower. Reading these findings leads to the realization that the habitual in home care work, and also in professional care work, would be relevant aligned focuses that are currently lacking in research.

In the future, a potential research agenda might give critical consideration to how affect as a force provides endurance for both wanted or unwanted habits in individuals and groups and their environments through multiple scales. Moreover, it might consider how affects can be manipulated or generated as a positive force to help create 'comfort zones' for bodies to act in certain ways, so that particular habits to grow, helping shape conscious thoughts, feelings and identities, as well as reproducing cultural practices important to positive health (Lea et al., 2015). The idea here is that, for researchers and policy makers, intervening in environments to change habits might be just as successful as the more traditional approach of providing logical arguments or rules that appeal to reason only (these being methodological challenges taken up later in Chapter 8).

Ordinary and everyday

NRT is concerned with spatial events that occur everywhere and all the time. Indeed, not aiming to be elitist, special or specialized (nor focus only on the elite, special or specialized), it concerns itself with the ordinary (i.e. non-remarkable) and everyday (i.e. common) events/practices and places in life (Andrews, 2014a, 2015). Ordinary and everyday events are often the regular things people do (e.g. walking, eating, washing etc.) that help them live successfully (Cadman, 2009). The ordinary and everyday places where they occur are inhabited or moved through regularly (e.g. rooms, sidewalks etc.). These happenings often remain outside of people's full consciousness although, being affective, they are far from neutral or trivial backgrounds, creating the feel and underlying rhythms of people's lives. Collectively creating the feel and underlying rhythms of homes, streets, towns, cities, regions and a globalized world (Andrews, 2014a). As Cadman (2009) suggests, the ordinary

Figure 5.4 Ordinary and everyday

and everyday are also transformative being the basis for much of life's enchantments, improvisations and enterprises.

Ordinary and everyday health geographies

Although there are overlaps, two main interests exist in this field of health geography: everyday home space and everyday public space. With regard to the former, Tucker's (2010a) aforementioned study, for example, conceptualizes home space as a central everyday resource for people with mental distress and illness providing regular and predictable activity; a 'mode of being' that acts as a reliable base to their lives. Other research here, meanwhile, is focused on particular ordinary and everyday activities that occur in home space that are more important than one might initially think. Notably, Anderson's (2006) aforementioned study, for example, considers the affective practice of listening to recorded music at home to enact forms of hope. One of his respondents had recently been made redundant, had been forced to relocate and now lived alone. He comments on his situation:

> ... [J]ust been a bit bored and lonely ... everything's closed around here ... I'm not doing anything at the moment... Sometimes I don't have the energy to do much else ... just sit here ... sit here too much and watch the world go past and see how shit the neighbourhood has become ... too many boarded up houses ... or go out and it's the same.
>
> (2006:742)

The respondent then puts some music on – Radiohead – as the interview progresses. He continues:

> ... I listen to this album ... in the morning and err ... if I'm feeling ... you know, like if I'm a bit low ... this'll cheer me up ... get me a bit more hopeful again. I don't know, it's quite melancholy ... but it's really beautiful ... and it stirs up emotion ... and I find it quite difficult to cry as well, and you need to at time to time ... so ... I ... this helps, so if I'm a bit low ... or I'm just lying here ... this is the song to put on _ umm.
>
> (2006:742)

Indeed, Anderson reflects that beyond the respondent talking about the uplifting affective quality of the music, here in the very moment that he was talking, the music created a new space-time – that escaped the old one – in which hope was able to more easily exist. It created an interlude; an intensified and different connection with life. A glimmer or spark of an alternative way of being where the respondent was momentarily disentangled from his everyday negative affects and emotions.

Meanwhile, still partially on the topic of music, Ravn and Duff's (2015) study of recreational drug use in private house parties described these common but hard-to-access and hard-to-map places. Using a map of their home – with locators 1–8 – he talked to two female respondents about their activities in the space:

MODERATOR: What about the bedroom, then?
RESPONDENT 1: I would say that [after dancing] you end up back on the couch again [number 6 on the map].
MODERATOR: Back on the couch.
RESPONDENT 1: Yes, 6 and then 7 [the bedroom] (both laugh).
RESPONDENT 1: If you are a little blazed (very intoxicated), then you just lie there [on the couch] and relax.
RESPONDENT 2: You can lie under the coffee table as well (laughs), if it's vacant.

RESPONDENT 1: Depending on where there is room for you, right, some of the places may get a little crowded.
MODERATOR: But is it when –
RESPONDENT 1: – you don't necessarily sleep.
MODERATOR: You don't necessarily sleep in the bedroom? You just go there to relax – or you go there to kiss with someone or?
RESPONDENT 1: – that as well.
RESPONDENT 2: If you have found a nice guy (laughs).
RESPONDENT 1: If you are just a little out [intoxicated], then you are still having a nice time with the others [in the living room], still keeping an eye on what they are doing, but your body is relaxed. Because your body can't take anymore, but your brain wants to go on (laughs). (2015:128)

With regard to the latter empirical interest (everyday public space), Andrews et al.'s (2013) study of CAM, for example, talks about how the actor-networks in this form of medicine spread out far beyond clinics (see also Lea et al., 2015). From the therapists' point of view, their practices helped encourage an holistic lifestyle enacted throughout their clients everyday lives (in homes, work-spaces and other places) involving continued belief in certain constructions but also regular body conducts. A respondent in their study comments on this approach:

> They [clients] need to eat, act, approach their thoughts and feelings and others in the way we discussed in [hypnotherapy and counselling] sessions, when they leave. The thing with complementary medicine is that it not just about the clinic and here. Its about everywhere they go and everything they do. An approach to life, otherwise it's not effective.
>
> (2013:103)

In another study in this area, also focused on wellbeing, Simpson (2014) considers the role of street music (busking) in the production of convivial 'healthy' public spaces: hospitable, sociable atmospheres that personalize sometimes bland commuter and shopping spaces and encourage interpersonal interactions – or at the least communal feelings – amongst strangers. Indeed, Simpson comments that instrumental sounds – by guitars, drums, violins, harmonicas – passed by on daily routines might annoy or offend some members of the public, but equally might elevate others and be an uplifting soundtrack to part of their daily lives.

Alternatively, on the opposite end of the spectrum of everyday wellbeing experiences, Bissell's (2010) aforementioned study thinks about pain and how, because it impacts relationships beyond the single body, it is part of everyday life and space. Bissell explains, how, as a force, pain constantly presses through the body, demanding physical responses ranging from subtle body positionings to full daily routines and rituals. Moreover, he explains how pain is amplified or reduced by particular environments which, for sufferers, might be more or less physically and/or emotionally comfortable. In sum, then, Bissell argues that pain involves complex distributed agency between internal human and external non-human worlds; pain flowing outwards into everyday spaces.

In sum, the styles and priorities of more-than-representational health geographies are now well-established, providing guidance and a firm foundation for future scholarship. There exists plenty of opportunity to broaden the horizons of current research and consider many other areas of health and health care.

6   Rethinking movement in health geography

From a change of location to movement-space

This chapter deals with the question: if NRT is primarily concerned with the immediate and happening in life, what then most fundamentally is it looking at? In other words, what, if anything, underlies all of the concepts and concerns of NRT outlined in the previous five chapters? The answer, as we shall see, is movement – specifically 'movement-space' (see Merriman, 2012) – which each concept and concern involves and requires. Before we come to the idea of movement-space, however, the first half of this chapter reviews health geography's traditional understandings of, and engagements with, space, place and movement across a number of empirical fields of sub-disciplinary inquiry. The idea of movement-space is then launched through a brief but powerful critique of these traditional understandings and engagements. The chapter closes by examining the idea of 'flow' as it appears in more-than-representational health geographies. Flow is particularly important because it is central to movement-space and many of its qualities which are explored in detail in Chapter 7.

Space and place in medical and health geography: mainstream perspectives

Space as a template, place as location

As noted in Chapter 2, in the 1950s medical geography emerged as a distinct sub-discipline of human geography. The wider quantitative revolution or 'empirical turn' to positivistic spatial science in the parent discipline was a key to this development at the time; a turn which involved looking for spatial patterns and relationships in collective human existence – some of its fundamental 'geometries'. Thus, in line with this priority, medical geography involved searching for spatial patterns and relationships in disease, health and health care. Within this project, space was visualized as a featureless, neutral underlying surface or template on which disease, health and health care phenomena could be located and quantified at particular locations. This process of location and quantification then allowed space to become mathematically distinguishable and dividable, thus representing substantial features of, and challenges in, human health. On one level, *at* places, rates, volumes and other localized measures became visible and calculable (such as levels of morbidity and mortality). On another level, *between* places, times, distances, movements and differences became visible and calculable (such as the spread of disease).

Spatial science was not, and is not, however, an unreflective mapping exercise concerned only with the measurement, calculation and spread of things. Indeed, most research within the tradition has been informed by specific theoretical traditions and perspectives that guide and frame inquiries and help interpret findings, notably early medical geography finding particular guidance in disease ecology (May, 1959). As Oppong and Harold (2010) describe, disease ecology provided, and still provides, explanations for spatio-temporal patterns of diseases by studying the factors and principles that influence these, its underlying assumption being that pathogens can only be fully understood if considered in the context of the locations in which they are found, and spaces through which they move. On the other hand, political economy has also been quite insightful, particularly informing research which describes and explains distributive features in the supply of health-care services (over international, national, regional and localized areas), and their relationships to usage, health patterns and health outcomes (Joseph and Phillips, 1984; Eyles, 1990; Mohan, 1998). The involvement of political economy has led to this field of research – the 'geography of health care' – contributing directly to mainstream health service and academic debates on rationing, efficiency and equity in service planning and provision.

In the last two decades, even though place is still essentially thought of as a location in much medical geography, a far greater appreciation has emerged in scholarship of what is found in and is influential in places, and of the complexity of the processes involved. This has led to much more sophisticated understandings and analysis, nowhere more than in research focused on the social determinants of health. For example, as suggested in Chapter 2, for a substantial period in the 1980s and 1990s debates existed as to whether health is more greatly influenced by the characteristics of local people (social composition), or by the services and facilities available to them (social context) (Curtis and Jones, 1998; Ecob and Macintyre, 2000) or by their social and cultural norms and collective practices (social cohesion and capital) (Macintyre et al., 2002; Mohan et al., 2005; Veenstra et al., 2005).

Place as consciously socially constructed

Two decades ago a new era of research was initiated that ultimately transformed the sub-discipline of medical geography into health geography, and changing understandings of place lie at the heart of this. Building on earlier commentaries and research on topics such as sense of place (Eyles, 1985), perceptions of health (Eyles and Donovan, 1986) and culture and health (Gesler,

1992), Kearns (1993) articulated a number of concerns with the reduction of places in much research to a mere points or dots on maps. Noting theoretical developments in the new cultural geography of the period, Kearns thus called for geographers to incorporate a dual 'place-sensitive', 'post-medical' perspective into their scholarship (Kearns, 1993). With regard to the former, he meant moving beyond previous preoccupations with distributional aspects of disease and medicine by examining the meaning and significance of places (re-imagined as social and cultural constructions) in health and health care. With regard to the latter, he meant problematizing medical categorization, assumptions and power and, moving beyond pathology, realizing health as both a mental and a physical state of wellbeing. Following Kearns' arguments – and some initial skepticism (Mayer and Meade, 1994; Paul, 1994), complementary advice (Dorn and Laws, 1994) and further explanation (Kearns, 1994a, 1994b) – the understanding has since developed in geography that health and health care are deeply affected by places and the ways in which places are reacted to, felt, felt about, and represented.

Specifically, an understanding has developed that, at one level, because of people and structural features (such as policies and technologies) in situ, places possess basic agency (e.g. hospitals provide institutional medicine, community clinics provide primary health care and so on). At another level, however, it is recognized that, underlying this basic agency, more intimate processes are at work whereby 'people make places' and 'places make people'. These processes start with the idea of embedded knowledge which, based on Heidegger's thinking, posits that humans can only relate to and beyond themselves through their situation, their literally 'being-in-the-world' and their relational consciousness of other humans and non-humans in the world (Eyles, 1985; Bender et al., 2010). Moreover, it is noted how forms of encounter with place allow their 'intentionality' and 'essences' to emerge. With respect to intentionality, it is posited that through human presence, perception and judgement, places themselves are ascribed meaning. From a phenomenological standpoint, just as objects' uses are critical to their meaning (i.e. objects are 'about' what humans do *with* them), so are places' uses critical to their meaning (i.e. places are 'about' what humans do *in* them) (Bender et al., 2010). With regard to essences, just as objects possess essences (i.e. their qualities that influence what humans feel emotionally *about* them), so do places (i.e. their qualities that influence what humans feel emotionally *about* and *in* them) (Bender et al., 2010). Humanistic writers explain that intentionality and essences result in individuals' feeling a sense of place (Relph, 1976; Tuan, 1977), whereby places can evoke a broad range of emotions, from basic to complex, and from extremely positive to extremely negative.

Armed with such ideas on places, a new generation of health geographers has sought to understand how places play a central role in a variety of processes in health and health care. As suggested in Chapter 2, the longstanding approach has been, and continues to be, to develop qualitative or mixed methods to observe in depth the health and illness lives of individuals and groups; to unlock the social and political structures imposed upon them and their experiences and agency within these (Kearns, 1993; Gesler and Kearns, 2002; Parr, 2003, 2004). Another has been to think of health landscapes as texts that in research terms can be read then decoded with particular methods and theories to unlock hidden processes (such as power and meaning) then (re)writen to represent these processes (Kearns and Barnett, 1997; Gesler and Kearns, 2002). Together these approaches constitute, and frame, many hundreds of academic papers and numerous books in health geography.

Emerging relational understandings of place

One criticism of the aforementioned understandings that has emerged in recent years is that places tend to be portrayed in research that adopts them as somewhat discrete and static phenomena; as stable, parochial centres of meaning resulting from social inscription (whilst little attention is paid to relationships that might exist beyond them) (Andrews et al., 2013). In response, 'relational thinking' complicates the conventional assumption that places capture intrinsic qualities. Simply put, relational thinking implies a twist in how place is theorized, evoking an image of places emerging not only in situ, but also through their connections within networks of translocal interactions. In other words, places are highly related to, and produced by, many other places through multiple scales. Indeed, relational thinking about place is usefully explained by Cummins et al. (2007), who emphasize that (i) whilst a conventional view of place focuses on fixed boundaries, a relational view focuses on fluid boundaries; (ii) whilst a conventional view of place focuses on content, a relational view focuses on influxes that change content; (iii) whilst a conventional view of place focuses on residence, a relational view focuses on mobility; (iv) whilst a conventional view of place focus on certain times and places, a relational view focuses on many and their change; (v) whilst a conventional view of place focuses on common understandings and constructions between individuals and groups, a relational view focuses on variable understandings between individuals and groups. Scholars have offered ideas for infusing relational notions of space and place into health research (Cummins et al., 2007), notably in the context of talking about complexity and complexity theory (Gatrell, 2005; Curtis and Riva, 2010a, 2010b). More generally, however, outside these dedicated commentaries, the emergence of 'relational health geography' is evidenced more subtly in the way that relationality has been used to frame a growing number of studies. These include, for example, considerations of health system processes and community outcomes (Durie and Wyatt, 2007), political ecologies of health (King, 2010), therapeutic, enabling and restorative environments (Curtis et al., 2009), and infectious disease and public health (Keil and Ali, 2007). Such works have enriched understandings of health and place by demonstrating the relational connections constituting health experiences. Most recently, however, health geographers have started to move beyond 'mapping' relationalities (i.e. articulating places as nodes in wider networks), and have embraced a further understanding that place is itself relationally performed, which is one basis for NRT (see also Chapter 3). As Anderson and Harrison (2010:16) state:

> It is not enough to simply assert that phenomena are relationally constituted or invoke the form of the network, rather it becomes necessary to think about the specificity and performative efficacy of different relations and different relational configurations.

Movement in medical and health geography: mainstream perspectives

The spread of infectious disease

Medical geography's longstanding concern for the spatial aspects of infectious disease includes a focus on movement, it being important on three levels. First, and perhaps most fundamentally, it is important in how disease spreads between bodies. Sabel et al. (2010), for example, describe how infection happens when a host body (typically a human body) is colonized by a foreign body (typically a micro-organism) that seeks to use the host's resources (typically energy and environment), moving into it to sustain itself and reproduce (almost always to the detriment of the host). Moreover, modes of transmission also involving movement arise in the guise of direct physical contact (between bodies) or indirect physical contact (such as with an unsterilized object), or through organisms being airborne, waterborne or vector-borne (a vector being any agent that transports the causal agent of a disease).

Second, movement is important to the wider spread of diseases and how, in terms of general prevalence, they move across space and throughout populations. Sabel et al. (2010) note how diffusion theory is used by medical geographers to work out the 'where and when' of the spread of disease. Regular contagious diffusion, for example, being spreads outwards smoothly relative to proximity, whilst heirarchical diffusion is spreads outwards with jumps occurring between larger population centres (often through direct transport networks). The spatial representation of this movement often comes in the form of GIS and/or investigative maps of association and understanding showing dot-point analysis, rates analysis and spatio-temporal dynamics (Rican and Salem, 2010).

Third, movement is important to medical geography in terms of other 'contextual movements' being contributing factors to the spread of disease. As Meade (1977) describes, underpinning disease ecology, for example, is the fundamental understanding that movements of people – at regional, national and international scales – are an important factor. Moreover, more recently, the insertion of political ecology into the theoretical mix has emphasized the importance of political, economic and social factors in this population movement. As Sabel et al. (2010) note, here scholars focus on population dynamics and attributes – four important subgroups of people being susceptibles (who do not have immunity), infectives (who are infectious), immunes (who have acquired immunity), and latents (who have been infected but who are not yet infectious). The movements, interactions and changing proportions of each group are recognized to be part of a disease's spatial progress.

Movements of health care workers

A reasonably sized body of geographical work considers movements of the health-care workforce, often at the collective level, and the social, political and economic forces that shape them at local (Brodie et al., 2005), national (Radcliffe, 1999; Courtney, 2005; Cho et al., 2014) and international and global scales (Kingma, 2006; Ross et al., 2005; George, 2015). One strand of research has considered for example (as suggested in Chapter 2), local supplies of labour, career decision-making (i.e. if and where to move to work), and the consequences of these decisions and actions for local communities (Barnett, 1988, 1991; Cutchin, 1997; Farmer et al., 2003; Laditka, 2004). Most of this research uses a political economy theoretical framework, and much of it links to global health contexts and concerns.

Movement as a social determinant of health

Elsewhere health geography has also engaged with human movement as a social determinant of health. This approach is common across four substantive fields of inquiry, the first being focused on international migration, where individuals and groups of people move between countries and continents with implications for their health and health care systems (see Bentham, 1988; Brimblecombe et al., 2000; Boyle, 2004). This research is not focused on the act of moving long distances per se as much as on the impact and consequence of the completed/past movements, a particular focus here being, for example, on ageing: on the migration destinations, motivations, decisions and experiences of older people and the implications for services and policy (Joseph and Cloutier, 1991; Moore and Pacey, 2004; Oliver, 2007). Recently, scholarship here has been augmented with far more critical considerations of the human lifecourse as played out, and moving, in space and time; studies being more sensitive to life's 'journey' including multiple moves and their meaning (Bailey, 2009; Jarvis et al., 2011; Schwanen et al., 2012).

The second field of study is on physical activity and health (e.g. Witten et al. 2008; Matthews et al. 2009). Building on an established sub-disciplinary tradition of measuring and mapping 'activity space' – i.e. the extent of space travelled in daily lives (Zenk et al., 2011; Kwan, 2012) – most recently the concept of walkability has emerged in the literature. Here, studies have focused practically on the production and form of walkable city environments (Ewing and Handy, 2009; Gehl, 2010), have considered potential social and structural facilitators of, and barriers to, walkability (Townshend and Lake, 2009;

Brown et al., 2009; Pouliou and Elliott, 2010), discussed methods to measure walkability and/or have actually measured walking rates and distances (Lin and Moudon, 2010; Giles-Corti et al., 2011), and have quantified and generalized public perceptions of walkability (Loukaitou-Sideris, 2006; Weinstein Agrawal et al., 2008). Within this particular scholarship – research on activity space notwithstanding – it is often the potential of movement, rather than movement itself, that motivates and surfaces in studies (Andrews et al., 2012).

The third field of inquiry where human movement is considered to be a loose determinant of health and wellbeing, is on the distances between health and social care providers and those in need. Incorporated into health geography under the general theme of accessibility and utilization (Joseph and Phillips, 1984), often highlighted in this literature is a distance decay in use (increasing distance = lower usage) and the concurrent impacts upon population health. Focused then as much on 'non-movements' as movements, an interest for example has been on older peoples' proximity to centralized facilities, to their children and to other carers who live away from them (Joseph and Hallman, 1998; Nemet and Bailey, 2000).

The fourth field of inquiry is focused broadly on transportation and health. As Widener and Hatzopoulou (2016) outline, diverse topics include the locations and routes of ambulances (Church and Meadows, 1979), helicopter/ambulances and emergency recovery (Schuurman et al., 2009), travel times and distances to hospitals (McLafferty, 2003), pollution from cars (Jerrett et al., 2005) and road safety (Whitelegg, 1987; Edwards, 1996). Here then, the particular mode of movement is the important factor in health and health care.

Qualitative/critical engagements with movement, health and wellbeing

Elsewhere across health geography – and occasionally social geography – human movement has been engaged more directly and critically by three substantive fields of inquiry that use predominantly qualitative methodologies. The first is focused on impairment and disability and mobility in these contexts; the (non)movements, challenges and (non)acceptance of individuals affected (Golledge, 1993; Gleeson, 1999; Crooks et al., 2008). A particular demographic focus in this literature includes, for example, the range and scope of older peoples' mobility (Meyer and Speare, 1985; Schwanen and Paez, 2010), their unique urban and rural challenges (Mattson, 2010) and policy implications (Mercado et al., 2010). The second area is a focus on fitness activities, much of this research unpacking the meaning and identity of fit bodily movement, the places where it occurs, and the power relationships at play (Bale, 2004; Andrews and Andrews, 2003; Vertinsky and Bale, 2004). Occasional foci here include, for example, attention to active leisure and lifestyles (Mansvelt, 1997) and gym culture (Andrews et al., 2005).

The third area – which being more of a broad approach does incorporate some of the research in the first and second areas – can be broadly described as 'new mobilities'. Attempting to escape the 'sedentariness' of much social science, it considers mobility as something in itself (as opposed to something that merely facilitates endpoints), and something worthy of academic attention itself. Thus it conveys the form, experience and implications of mobility (its motivations, meanings, structures and relationships to a globalized world), particularly in the contexts of residential relocation, transport, leisure, treatment and work (Gatrell, 2011). Specific focuses here for example are on medical tourism (Crooks et al., 2011; Snyder et al., 2011; Kingsbury et al., 2012) tourism mobilities and wellbeing (Williams et al., 2000; Schwanen and Ziegler, 2011) and movement during mental health crisis (McGrath and Reavey, 2015).

Towards movement-space

> [H]uman life is based on and in movement... movement captures the animic flux of life.
>
> (Thrift, 2008:5)

Two critiques are possible of the research on movement outlined in the previous sections, based around the observation that, despite the considerable breadth of empirical interests, movement remains rather narrowly defined and only partially conveyed. First, that obvious forms of human movement – such as migrating, walking and working out in gyms – do not have to be the only subject of study. All life requires movement in every moment; all bodies and most objects move to some degree, from the motions made on the very smallest of scales and in the briefest of times, to the infinite ways individuals and societies go about their daily lives. Second, that it is possible, and arguably necessary, to go beyond current research and get a far better grip on the act of movement itself. Whilst currently, quantitative studies show general trends in collective macro-scale movements over long timeframes, qualitative studies are concerned, for the most part, with what it means to move or not move, and/or the power relationships involved. Neither group, really engages with the processual elements of movement; how movement happens, appears and feels in the moment. Rectifying the situation necessarily involves thinking about particular qualities of movement as expressions in their own right, with their own specific character and consequences. Indeed, NRT is arguably well placed to shed light on the fundamental making, immediacy, physicality and feel of movement so that movement might be studied as something in itself; in slightly crass terms, becoming a 'unit' of analysis in health geography.

Debates on the nature of space and time are longstanding and fundamental in human geography, and necessarily involve consideration of the character and production of movement (Seamon, 1979; Harvey, 1990; Massey, 1992; Doel, 1996; Thrift, 2003, 2004c; Miller, 2005; May and Thrift, 2003; Jones, 2009; Merriman et al., 2012). Nevertheless, as Thrift (2008) argues, movement is the leitmotif of NRT, and perhaps its most central and fundamental underlying concept (as suggested above, NRT recognizing that everything in life involves and depends on some kind of movement, even if as a counterpoint or contrast). NRT recognizes that movement is more expansive than, for example, humanistic concepts such as belonging or meaning, which fall under quite specific emotional categories (or are very quickly bypassed in studies in a search for their construction/making). (Thrift, 2008). In NRT's studies on movement, dance has been a particularly popular empirical topic that, as McCormack (2008) notes, has been used to illustrate how, through 'techniques of the body' – such as demanding space, and showing different spacings and shapes – bodies move in many different ways: spatio-temporally, kinaesthetically, affectively, aesthetically, collectively, culturally, politically and imaginatively. Thus, how they have the power to write and generate different kinds of space-time, enticing and seducing in a permanently disappearing act (see Thrift, 1997; McCormack, 2003, 2013; Dewsbury, 2011). Building on this literature and the array of spatial thinking on movement emanating from NRT more generally, is Merriman's idea of 'movement-space' that seeks to augment the notion of space-time being a more useful concept for researching social contexts geographically (Merriman, 2012). In this key work Merriman tells a brief history of space and time in geographical thought, tracing a fundamental agreement across many eras of scholarship, originating in Newtonian thinking, that events unfold *in* space and time. He reviews substantive traditions including spatial science (with its treatment of absolute space and its privilege over time), Hägerstrand's time geography (the mapping of routine human behaviour) and the humanistic and social constructionist emphasis on micro-scale social space (structures of feeling). Merriman notes how, in recent years, a realization has occurred in geography that not only does the enactment of events create individual space-times, one might think about 'parcels' of space-time emerging from them (particularly, for example, in the theory produced by scholars such as Thrift, Massey and Harvey and in the empirical work on dance noted above). Merriman however challenges the notion of space-time here and questions why it is still privileged and regarded as foundational even in this latest radical thinking. He suggests that even this highly relational view of space-time reflects the domination of western science and philosophy which can lead to crude physicalism and convenient non-reflective labelling ('space-time' being used loosely across numerous contexts). Reaching for a solution, Merriman draws on Bergson's and other philosophical ideas to inform his idea of movement-space. Merriman's movement-space incorporates space and time but considers them to be no more important than a range of other qualities that speak equally to the progressing onflow of existence and which, by dealing directly with the realms of force, energy and sensation, speak even better to the nature of human practice and experience. Specifically, Merriman (2012) asks: 'why not position movement, rhythm, force, energy, or affect as primitives or registers that may be of equal importance [as space and time] when understanding the unfolding of events?' (2012:24).

Although I support much of what Merriman proposes, I argue however that there might not necessarily be an either/or decision to be made between space-time (as used in this book's title) and movement-space (as examined in this and the next chapter). The term 'space-time' seems to suggest something that necessarily exists within various consecutive measures, that fold onto it (thus make it). In other words, movement-space is quality-laden space-time; movement-space describing performances that create space-time and the body's openness to it. This point is supported a number of commentators such as Tucker (2010b), who posits that an NRT or Deleuzian perspective is that space does not pre-exist, and humans do not operate within it, but that space-time is reproduced and re-enacted by the distributive movements of bodies and objects that create it (territories being made and unmade by these flows that become more or less familiar and dominant). In other words, NRT thinks of space in topological rather than topographical terms. Thinking about the conditions of space's emergence, it recognizes space as a socially produced set of happenings activated by movements within an assemblage (Thrift, 2000; Boyd, 2017). Or, in short, as McCormack (2008:1823) argues: 'spaces are – at least in part – what moving bodies do'.

Chapter 7 explores some of the consecutive measures or qualities of movement-space. To close this chapter, however, the idea of 'flow' is introduced – it being a general experience and sensation of movement-space – and how it arises in more-than-representational health geographies.

The example of 'flow' in more-than-representational health geographies

A scientific understanding of flow is a steady continuous, cohesive stream of movement of liquid or gas or solid, measured moving past a given point in a given unit of time (e.g. litres per second). The idea of flow has however also been applied to the social world, with regard to the collective flows – in terms of ease of motion – of humans and/or their technologies/products within and between buildings, neighbourhoods, towns, cities, regions and counties. On an experiential level it is also the sensation of moving (of flowing) with or against these flows.

As suggested in Chapter 3 (in the earlier discussion of onflow as the initial moment of all flows), there have long been calls in human geography to consider, more critical and directly, the nature, character and experience of the flows of the social world across space, as a way of escaping the production of rather static representations by much contemporary social science inquiry, and to account for the nature and increasing prominence of movement in an ever globalizing and technologically driven world (Shields, 1997). These calls were recognized by the emerging new mobilities paradigm in human geography which, as we know, considers mobility as a thing unto itself, worthy of academic attention itself (rather than its endpoints), and the structures and experiences involved (Sheller and Urry, 2006; Gatrell, 2011;

Cresswell and Merriman, 2011). As Shields (1997) points out in an introduction to a collection he edited:

> [T]he notion of 'flow' most widely known from the work of Deleuze and Irigaray occurs repeatedly in social theory. Associated with a paradigm shift within cultural studies and sociology from the analysis of objects to processes, it is also linked by geographers to the notion of 'nomadism' and the breakdown of the fixity of boundaries and barriers.... In effect, the dominant metaphors for discussions of sociality have swung from models of affinity to those of viscosity ... Flows are spatial, temporal — but above all, material. In this issue, we advocate an analysis of flows, which examines their qualities, but avoids their analytic reduction to causes, origins and destinations. Final effects, like originary causes are in the end irretrievable and irreducible. The ambiguities of between-ness, edge-states and borderline conditions, of the effects of cinema's blur of images, are some of the host conditions of flow.
>
> (1997:2)

In health geography flow is understood as a way through which, and where, health happens; one field of research being focused on how flows are viewed, performed and felt in health contexts (Gatrell, 2011). Gatrell (2013) for example considers the therapeutic qualities of the flow involved in the practice and experience of walking from one place to another. Spinney (2006), on the other hand, adds further ideas through a study of cycling, arguing that their flows through place help define cyclists' engagements and understandings of it, not only because they are always differently relational to that place as they move but because the kinaesthetic sensations of that movement – such as rhythm, effort – are important to knowing place. Indeed, in his paper Spinney paper wistfully reverses the classic humanistic notion of 'sense of place' to a 'place of sense'. In another study Evans (2010) observes how so much research on obesity, in addition to uncritically accepting public health imperatives, describes relatively static geographies (for example, where the fattest people are located in cities); people being represented at fixed points within bounded spaces. Moreover, she argues that central to this research are ideas from nutritional sciences and public health about the balance between energy consumed and energy expended (this typically being aspatial in the way that it is measured on an individual basis). Evans suggests however that, in contrast to what most research shows us, the reality is that the human body is always in motion, always changing, always subjected to varying messages and desires. Evans thus posits that surely fat bodies own flows, and the way in which they are included in or excluded from dominant social flows are important subjects for geographers to consider?

Other research in health geography is particularly concerned with exploring the constant interplays between the body, landscape and the mind during the flow of activities; how emotions, feelings and movements work together in ways that are not always consciously stated and cannot easily be captured by language and text (Greenhough and Roe, 2011). McCormack's (2003) aforementioned study of dance movement therapy, for example, demonstrates how ethics and emotion can be expressed through movement – the flow of dance – and its affectual registers (including in touch and gesture). Dance, according to McCormack, involves sense-making without necessarily involving contemplative cognition. In another study, Barnfield (2016), in the context of recreational running as a public health intervention, suggests that movement itself (involving flows, spatial processes, skills and efforts) provides 'excessive' experiences involving many of the body's registers at once. Indeed, Barnfield suggests that the body both moves and feels at the same time, the two being closely bound. In other words, the body moves as it feels, the body feels as it moves, and the body feels itself moving (a phrase extended from Barnfield, 2016, and Massumi, 2002). Finally, Pitt (2014) also considers flow in the context of community gardening. For Pitt, flow involves regular and loose movement but is also something more: a less-than-fully conscious feeling state of effortless absorption in this movement with positive consequences for the nature of participation. She argues that those participants in her study who were able to enter the flow of the activity in body and mind benefited far more from the opportunity and obtained most therapeutic affect; they finding tranquillity, escape, mental stillness and relaxation in mundane repetitive movement and actions. One participant, whilst finding digging physically challenging, also found it:

> mentally quite relaxing [...] nice, tranquil, easy and err you're just there doing it with your thoughts to yourself.
>
> (2014:87)

This experience of flow is somewhat similar to what occasionally in sports/ fitness geography is described as a 'flow state' (what in lay terms is known as 'being in the zone'). A feeling state of absorption and strong unselfconscious participation that reduces a participant's thinking time, helping them to achieve a higher level of performance with less conscious effort (flow thus being both in their bodies and with the activity). Indeed, a recent review of sports and fitness geographies Andrews (2017a), argues that in Deleuzian terms these kinds of flows can also involve immersion in 'crystal time' (or crystals of time), a patterned energy and intensity that is not necessarily mapable onto linear, artificially segmented clock time (Deleuze, 1989; Glennie and Thrift, 2009). The paper explains:

> Being more typical of the human experience, crystal time/crystals of time is lived time that flows along, 'stretching' when time seems to slow (for example when a participant in a fitness activity wants an event to end), 'compacting' when time seems to speed up (for example when a

participant in a fitness activity needs more time to compete and does not want an event to end) or 'shallowing' when time seems to matter less (for example when a participant in a fitness activity is intensely occupied in a performance).

(Andrews, 2017a)

As Andrews suggests, in terms of experience, although participants in fitness activities might lose themselves in the movement refrains that constitute the flows of their performance, regularly or irregularly intervening moments of formal structure interrupt periodically (such as a whistle), announcing a sudden conscious return to clock time.

7 Qualities of movement-spaces in more-than-representational health geographies

As suggested in the previous chapter, the qualities of movement-space are varied (including affect, energy, vitality etc., addressed earlier in the book). This chapter explores those most associated with its physical and felt forwards progression/advancement and also its aesthetics: those that perhaps speak most to what fundamentally movement-space does, namely speed, rhythm, momentum, imminence, encounter and the immediate contrast of stillness. For each quality attention is given to popular/lay, scientific and social scientific understandings, to how NRT might instead view it, then finally to how it has arisen in explanations of health in health geography and particularly in more-than-representational health geographies.

Speed

A scientific understanding of speed is the rate of motion of an object through space measured in terms of how much distance is covered by it in a set amount of time (such as metres per second or miles per hour). A scalar quality, speed can be described numerically by its magnitude alone (velocity, on the other hand, being a vector quality, says something more; it being the speed of an object in a given direction). Only massless photons of light immediately possess their maximum speed, otherwise any object possessing mass will necessarily have to accelerate to reach a certain speed. Once an object is in motion, its speed can be constant or increasing (accelerating) or decreasing (decelerating), and this might happen either uniformly or irregularly over a given duration. Human subjective judgements become involved when the speed or velocity of an object are compared with those of other objects, this relationality providing subjective descriptors and attributes (such as fast and slow). Indeed, in the social world speed, and these attributes, are also an aim (to go fast or slow), a practice (going fast or slow) and a sensation (feeling fast or slow). Moreover, involving many objects moving in multiple related trajectories, are complex events and subjectivities such as tasks judged to be done quickly (with speed) or slowly (lacking speed), and numerous collective human creations and structures which have their own speeds (such as market development, policy formulation and implementation, manufacturing production/output, building construction, consumer purchasing and spending and so on). In practice, because life necessarily has movement, all life possesses some speed. Certain objects important to human culture – such as monuments – might not move, but humans and other objects move in relation to them (see 'Stillness', p. 135), and even within seemingly static living organisms, internal and/or minute movements possess speed.

Speed has emerged as a new concern in critical human geography and the social sciences more broadly, particularly in terms of how societal economic and technological developments have led to, and characterize, the 'speeding up' of the world (Latham, 2008; Bergmann, 2008). Whilst patterns and cycles of media, work, politics, construction, communication and consumption have all speeded up, these are accompanied by an accelerated pace and motion in everyday life, what Levine (2005) conceptualizes as an increasing 'busyness' (that he describes as speed + activity). This busyness, although generally present, differs between individuals and cultures, and is variously an experience, a way of life, a strategy, and an individual and social statement. Examining this diversity further, scholars have argued that although pace of life and sense of time differ between places and cultures (including differences in timekeeping and expectations of promptness), one facet of an increasingly globalized world is a 'multitemporal society' where humans experience, and often travel between, different paces of life and senses of time (Levine, 2008) (it also recognized that the speed and form of this travel itself have also become an important part of the human experience; Bergmann, 2008). Meanwhile, moving down in scale, there is a realization that whilst societal development generally leads to an accelerated pace of life, this is uneven and not applicable equally to everyone in a specific place (such as a city or neighbourhood). Indeed, for some people, although they might be subjected to the very same development processes that generally speeds things up in their locality, the outcome for them might be that their pace of life slows (Hubbard and Lilley, 2004), whilst other people might prefer for their pace of life to be slower and might actively seek to make it that way for themselves (Pink, 2008). Moreover, in terms of health, there is a recognition that, whilst people often claim to be happy with their increasing pace of life (because, for example, of the personal fulfillment in exerting effort, the financial rewards, increased consumption and socialization; much of this being arousal of various kinds), paradoxically it can also involve increases in stress and poor health outcomes (Garhammer, 2002). Beyond this research on broad social trends, the aim of NRT is to take the next step and show the immediacy and sensation of speed itself. To show its uses and consequences in its raw form, often as it occurs at the micro scale in bodies and objects. In this way NRT relates to, and has as much in common with, the aforementioned scientific understandings of speed.

Figure 7.1 Speed

Speed in health geographies

Traditional medical and health geography has developed a rudimentary engagement with speed as it occurs across space and of/in place. This arises, for example, in a positivistic research tradition where time-series and spatio-temporal analysis is applied to a range of phenomena (Meade and Emch, 2010) including the spread of diseases (Smallman-Raynor and Cliff, 1991; Gould, 1993; Rogers and Randolph, 2000), and treatments/hospitalizations for specific conditions (Crighton et al., 2001, 2008; Leong et al., 2006). This research presents – often with detailed computer-aided illustration/ GIS – a sequenced progression of phenomena collectively occurring over macro scales. Otherwise in the political economy tradition there is a focus on the speed of market developments, such as expansion and contraction of private care businesses and sectors in given areas subject to new conditions (Phillips and Vincent, 1986; Andrews and Phillips, 2002; Ford and Smith, 2008).

There is however a more direct engagement with the motion and sensation of speed in more-than-representational health geographies. On one level there is an emphasis here on speed as it relates to progress. Duff (2014a), for example, uses the case of mental health recovery and spaces of drug use to illustrate his theoretical points about how the process of becoming healthy is accelerated/speeded up or decelerated/slowed by particular assemblages. Indeed, here speed is not necessarily of any one particular body or object, but of an overall process, yet even this process does have physical elements that move (i.e. speed within the assemblage itself – there being speed to the bodies, objects, forces, affects and relations involved).

Another engagement with speed in more-than-representational health geographies focuses on the aforementioned increases in the pace of modern life and their health consequences. Here there has been some consideration of the role of 'slowness' in interventions that aim to provide a break from everyday life and its stressors. In Conradson's (2005) work on rural respite care centres, for example, respondents used the service and facilities as a place and opportunity to slow down, re-centre and reflect upon their lives. Similarly, Conradson's (2007) aforementioned study of therapeutic retreats argues that these places purposefully provide a slower pace of life; in terms of perception, time within them appearing to guests to have slowed since they entered (akin to the idea of felt 'crystals of time' as discussed in Chapter 6; a patterned energy and intensity felt as duration as opposed to clock time).

As Andrews et al. (2014) highlights, however, moments of wellbeing that escape the intensity of everyday life do not necessarily have to be slow. In this research, a personal moment was articulated by one of the research team in their field diary, where speed and motion was important to them and also their relationality to other bodies and objects. They describe an everyday driving experience, and how as an affective feeling state flows into more fully conscious emotional contemplations:

> I'm sitting in my car driving home. The road is clear, the car is at a comfortable temperature, the weather outside is fine. I look ahead at the road through a clean and spotless windshield under a clear blue sky. I absorb the bright and colourful surroundings passing by; the street lights, brick homes, cars, trees, and stores of the suburban neighbourhoods. There is no wind, the trees are still. The traffic is smooth and moving freely, putting me completely at ease. Every car is travelling at the same speed, in sync with me and each other. All are moving together, like a pod. All is quiet, apart from the rumble of my tires, I hear the giggles of my son engrossed in his video game behind me. The sound of the video game is a low hum from a racing car, brrrrrrrrm, brrrm, brrrrrrrrrrrrrrrrrrrrm. My footing is light on the accelerator, my mind

is open and my heart is content as I enjoy the relaxation, the feeling of the moment, of being mobile. Then, I look to the right and notice a Gas Station that I have driven past a thousand times before but have never paid attention to. I am filled with fond memories of another city I used to live in. I remember a similar Gas Station there that I went to every morning to buy coffee. It was never too crowded but always filled with friendly customers enjoying small talk. It was brightly lit with a self-serve coffee counter carrying my favorite coffee. Two sugars, two cream, two dispensing buttons and voila, my coffee was made. The employees were happy to be there, always smiling, interested, and talkative. It was a great morning every morning in that place. I slowed the car down to stare at the gas station a bit longer. It had triggered memories and feelings as I had driven past...optimism, warmth, and acceptance.

(2014:215–216)

By far the most common research engagement with speed however in more-than-representational health geographies is how it occurs in particular movement/fitness activities and the senses and sensations involved. One group of studies conveys bodies propelling themselves at speed, relatively naturally and unassisted. Middleton (2009), for example, in her examination on the temporality and spatiality of urban walking, remarks that transport policies emphasize increasing speed with a view to reducing (clock) time taken for journeys. However, she posits that for walkers in particular, speed and time are experienced and practised differently; people for example becoming more acutely aware of their lack of speed and the length of time taken particularly when they are held up and have to wait (again akin to crystals of time). In this regard, one of Middleton's respondents comments on the tactics he employs to maintain his speed on his walk to work:

This time last week I had just come back from holiday and my traffic light awareness function was not quite operating at its optimal capacity. You get a sense of timing when doing the same route each day. Where to cross and lose least time. Which corners are blind and make the heart race. And which order the traffic lights change in so you're one pace ahead of the infrequent walkers. Last week my timing was all over the place – now I'm back.

(2009:1951)

In another study the specific natural propulsion considered is swimming. Foley (2015) considers sea swimming as an embodied, affective therapeutic activity. His respondents talk about speed being one sensation, amongst others, important in their overall experience. A respondent remarks:

I love it (swimming). It makes me feel very complete or something; very relaxed. It's quite ... buoyant actually. I love the sensation of weightlessness and the kind of speed and momentum you can get up in the water. There's a kind of a playful element to it as well I love.

(2015:476)

Another group of studies is similarly concerned with bodies propelling themselves forward yet, in addition to articulating the senses and sensations involved, they are focused on activities where specific and often complex technologies – such as bikes – are central to the speed attained and overall performance and experience. Cook and Edensor (2017), for example, in their ethnography of cycling, talk about speed and rhythm at different stages of a ride and training year, and also about the importance of sensing surfaces and textures in perceptions of speed (for example, how the perception of speed changes between smooth and bumpy roads and also between day and night):

I set off slowly. For me, this is a function of age. I used to set off at a rapid pace but now I let my body slowly unfold in the first 30 min of the ride and get into rhythm. I start to get into a rhythm after the first 15 min or after the first hill; whichever comes sooner.

(2017:13)

I am always relieved in the autumn that the racing season has ended. I ride in low gears and my pedal cadence is high. This is a rest time, one in which to regain suppleness, not to build strength. After Christmas, I begin to push myself harder. Larger gears, same pedal cadence if possible, therefore much faster. In spring and summer, I'm ready to compete. It's too late to change the body. All the hard work is done in the winter and early spring.

(2017:13)

At night everything is more intense. I feel as if I am going faster, an intense experience of the landscape produced by the beam. I don't see the landscape beyond the beam but I can feel it, rather like the presence of another. In the dark, rain enters my vision as silvery pelts; the handlebars a silhouette against the dark road below. I hear my own breathing, not labored. The well oiled chain smoothly turns over the cassette and chain rings. It rattles a little in the front mech if I stand on the pedals. The wind whistles through the straps of my helmet, when I go downhill.

(2017:14)

Likewise Brown's (2017) study of walkers and mountain bikers highlights how varied textures and surfaces are not only important experiences in themselves, but change the perception of speed. One of the respondents' comments suggests how perceived speed of movement in highly textured terrain can matter even more than actual speed. Again, the kinds of speed noted here helping create, in Deleuzian terms, the experience of felt duration:

[T]he manicured paths or wide [paths] ... you just get bored, time feels like it is passing really slowly ... once on the natural trails you probably take longer but it feels faster, and you feel you've been somewhere.

(2017:2)

One of the take-home messages of all this research on movement activities has to be that bodies enjoy speed on both conscious and less-than-fully conscious levels, this surely being an important consideration for public health planners and other officialdom. The same, as we shall see, applies to the other qualities of movement-space considered next.

Rhythm

Rhythmicity is crucial to the understanding of the temporality of flow and the role of flow in changing the commonsensical appearance of clock time.

(Shields, 1997:3)

Rhythm is understood scientifically as a segmented forwards movement defined by predictable timing and spacing; repeated intervals of either alternating strong and weak periods or repeated intervals of progressions each followed by complete stops. These intervals create a uniform pattern which provides a consistent pace to the rhythm. Specific types of rhythm include regular (constant interval lengths), progressive (interval lengths change, often building in noticeable steps) and flowing (changes occur consecutively up and down in interval lengths). Echoing the scientific understanding, but not as precise, are numerous human expressions and knowings of rhythm. Rhythm is, of course, most typically associated with the production and consumption of music. Rhythms might also occur in the economy, or more abstractly in the rolling out and spread of ideas and policies. Notably, in terms of immediate human participation, the timed spacing and movement of bodies and objects might also constitute physical rhythms in numerous areas of human existence. Thus, spatially one might experience and participate in the rhythms of our homes, workplaces, consumer spaces, neighbourhoods, cities and regions, and temporally in the rhythms of moments, hours, days, weeks and years. Humans participate in and experience rhythm through being immersed in it to some degree physically and mentally. Rhythm is sensed, registered and acted either consciously or less-than-fully consciously as the body becomes both in tune with, and part of, the rhythm; the regularity of the rhythm providing a comfort and incentive to move along at a certain pace with other bodies and objects, it feeling natural to do so (or the opposite being the case when it feels awkward, difficult and unnatural to be out of sync with the prevailing rhythm).

On an individual level then, as Merleau-Ponty (1962) suggests, humans inhabit rhythm: a movement being learned when the body has understood and internalized it, and has incorporated it into its world (Spinney, 2006). As McCormack (2002) suggests, this process is an important part of the conduct and experience of life and, as such, has attracted the attention of philosophers and social theorists for the best part of a century, particularly with regard to how rhythms define and collide. As part of his wider explanation of 'rhythmanalysis' Lefebvre (1992), for example, talks about how rhythm creates an animated space which amplifies the bodies involved, making them larger in terms of apprehension, particularly in relation to those observing. Moreover, for Bergson (1911a,1911b), the synchronization or conflict between co-existing social rhythms might lead to degrees of mutual relaxation or tension amongst bodies on different levels of consciousness. Ultimately then, as Deleuze and Guattari (1988) posit, rhythmic relations and expressions between social milieus create 'territories' of different style that are not hard barriers but motifs; counterpoints to one another.

In sum, rhythm is an essential part of successful human existence. Without the body's ability to act less-than-fully consciously with and within rhythms, every single aspect of everyday life would require pause and constant consideration (Spinney, 2006). On a collective level, less-than-fully conscious involvement in rhythms becomes an important form of repetitive organization, and a fundamental way in which humans inhabit and produce space. As Lefebvre (2004:15) suggests, 'everywhere where there is interaction between a place, a time and an expenditure of energy, there is rhythm'.

Rhythmic health geographies

In much traditional health geography, rhythms are evident in trends and representations of phenomena 'extracted' from the bigger picture. Demographic studies have, for example, conveyed and displayed the long-term spatial and temporal rhythms of population changes, as they themselves relate to the rhythms of population health and the challenges for governments and health sectors in particular places (Rosenberg and Moore, 1997; Smallman-Raynor and Phillips, 1999; Phillips, 2002). At the same time, qualitative research displays some consideration and representation of rhythm across a wide range of empirical subjects, fitting loosely into three broad categories. First, there are studies that convey the routinized, processional and rhythmic nature of treatments, caring practices and the places where they occur: for example, in nursing (Andrews and Shaw, 2008), addiction detoxification (DeVerteuil and Wilton, 2009; Wilton et al., 2014) and the routinized nature of informal and formal home and community care (Wiles, 2003). Second, there are studies that explore how the body becomes synchronized with broader rhythms that surround it as part of therapeutic places and experiences, such as rhythms of the sea in coastal life (Kearns and Collins, 2012), and rhythms of fitness activities and environments (Bale, 2004; Andrews, et al. 2005). Third are studies that convey the daily geographies in the lives of people with chronic conditions (Moss, 1997; Moss and Dyck, 2003) including mental illness (Dear and Wolch, 1987; Parr, 1998, 2008). These studies pay particular attention to the embodied rhythms – oftentimes part of makeshift survival strategies – that arise as peoples' health status and social contexts change (Rowles, 2000; Antoninetti and Garrett, 2012).

Despite their breadth, the above studies are however quite rudimentary and partial sub-disciplinary conveyances of rhythm. In contrast, more-than-representational health geographies get to grips far more with the compositions, performances and feelings of rhythms and their centrality in health experiences. Three empirical fields in particular stand out in the emerging literature: holistic practice, retreating, and movement activities. With regard to the first, Andrews et al.'s (2013) aforementioned study of CAM, for example, describes how therapists seek to create a rhythm to their therapy sessions. An extract from their field diary reads:

> Mike practices (chiropractic and massage) with a boundless energy. He moves quickly around his treatment space, busily but not erratically. He asks his client questions whilst moving a body part or inspecting it, listening to the answers which seem to shift him automatically to another question and to another body part. His session bounces along like this, rolling seamlessly from one question and body part to another question and body part. The client seems to realise their role, and play their part in maintaining the momentum, moving when subtly indicated like a partner in a dance. Only when Mike's session ends do I fully realise how energetic it was. Like the feeling when you get off a fair ground ride or motorbike and plant your feet firmly back on terra firma.
>
> (2013:106)

Focusing on a particular kind of CAM, Boyd (2017) describes the rhythms of dance movement therapy. In particular, she observes that in this modality, rhythm goes beyond sound heard and reacted to. She suggests that rhythm begs for a muscular and nervous response from the body which it then envelops, helping it to become in new ways. An extract from her field diary reads:

> One of my first forays into geographical fieldwork was an encounter with dance movement therapy and a private session with Swagata Bapat, who was once a professional Indian dancer and now a practitioner of 5rhythms™ dance therapy. Although it is not the only way to practise, she decided to dance a 'wave'–a continuous dance which moves through five rhythms of music: flowing, staccato, chaos, lyrical, and stillness. Unlike a facilitated class, when practised privately, the dancers choose their own music to match their understanding of the rhythms. Music that supports

'flowing' might be a slow ballad or an andante from a piece of classical music. Music that supports staccato might be a 'punchy' rock song. Music that supports the chaos rhythm often has a strong and rapid 'drum and bass' line, such as electronic music that has a 'tribal' quality. Lyrical involves a slowing down to a more moderate pace, with 'pop' style songs that are cheerful. Stillness makes use of very slow, droning, or melodic instrumental music.

(2017:76)

Elsewhere, research focuses on holistic practices that are embedded in everyday life. One of the participants in Lea et al.'s (2015) study of mindfulness meditation commented, for example, that as an everyday practice inserted into her daily routines, it helped disrupt her old unhealthy rhythms:

A lot of my life I've just been running from one thing to another like in a frantic kind of excited way because I'm quite an … energetic sort of person … I was doing lots of different things … But I feel like I'd just been running from one thing to another and never stopping. And so this maybe has given me a chance to stop.

(2015:55)

The subject of health practices that disrupt old routines is continued in the second theme in the literature: retreating. Here places/settings of retreat are known to introduce alternative and slower rhythms that provide opportunity for relaxation and contemplation. Conradson's (2005) study of rural respite care for carers of persons with physical impairments, shows how these settings allow people to remove themselves from the normal rhythms of their life to assist their wellbeing, a respondent in his study stating:

I look forward to it as a time of relaxation, of being just me. I look forward to it as a place where the pressure is lifted. Here I am away from my phone. I don't have to look at my diary. I can just be me. I can relax and enjoy just doing what I want to do. There is no pressure from anybody else. You take the lid off the pressure cooker, which then gives me the strength to go back and carry on.

(2005:8)

Similarly, Conradson's (2007) aforementioned study of therapeutic/spiritual retreats describes how these settings provide a break from the normal rhythms of life; a different and less intense rhythm. Conradson observes, for example, how the rhythm of the day was based around prayer, sung worship and shared meals.

With regard to the third theme, movement activities, a number of different forms have been considered. Barnfield (2016) suggests, for example, how recreational running requires a regular routine of training with its own weekly rhythms (to maintain regimes and underlying fitness), but is itself also rhythmic in the moment. In particular, Barnfield talks about synchronization and human company, about running drawing bodies together in relational closeness and intimacy. A club organizer in his study comments:

> We have learnt that we need to get the runners to either run a route that everyone can see each other or a couple of laps so people feel they are running together. It is important for people to feel like they are running together especially in the shorter distances. People need to be able to share experiences of running in a group, otherwise it is not good for people to get together. People have many options in Sofia so we need to show them this is an activity they can do, in short time, that is fun, and with other people.
>
> (2016:285–286)

(notably, in addition to Barnfield's work, other studies describe similar relationships between terrain and the rhythm of running; see Lorimer, 2012, and Cook et al., 2016). Elsewhere Middleton's (2009) examination of the temporalities of urban walking discusses how, in contrast to the rhetoric in transport policy based on times and distances, walkers experience the city differently, rhythm playing an important part. Moreover, she describes how rhythms are known to change during the course of a day, one of her respondents stating:

> I love walking through London at night.. you're allowed to just dawdle and piss about in a way you can't in the daytime.. in the daytime I'm always doing something whereas at night I'm relaxed.
>
> (2009:1957)

Further, Middleton describes how these rhythms are not only non-linear experiences of space and time, they can even facilitate, or be undertaken, through new modes of thinking. A respondent in her study comments in their diary:

> I often have a walk if I need to think about something and walk with partner at the weekend-for a couple of hours. Walking produces a metronome effect. Moving fast sometimes means that it is difficult to think something through. Moving slowly allows my mood to be more considered, plus therefore makes it easier to think.
>
> (2009:1956)

Other research, meanwhile, considers swimming. Foley (2015) for example discusses its therapeutic qualities, the respondents in his study commenting on the rhythm of the body in motion with the sea:

> But also when I am swimming along on my own, especially in the summertime, I have to say I enjoy it more. You just go into this lovely rhythm. I suppose it's a bit like meditating. You just go into this calm state and

you listen to your breathing and you visualise your body and what it's doing with each stroke. And if there's a slight wave you just become aware sort of, of your arm falling at a different ... falling into the water you know ... touching the water at a different ... you were expecting and oh ... in your head you reach it at the top of a wave and in fact you reach it at the bottom and it's very difficult to explain. It's just like you listen to your body, that's rather nice.

(2015:476)

The engagement with rhythms in movement activities has been extended in recent years by the emergence of new research at the disciplinary interface of health geography and sports geography; this for example illustrating spatial and temporal rhythms in diverse practices including kayaking (Waitt and Cook, 2007), golfing (Bissell, 2013), climbing (Barratt, 2011, 2012) and cycling (Spinney, 2006; Cook and Edensor, 2017). A relevant debate here, highlighted by Andrews (2017a), is 'synchronicity' or 'synchronization', which considers coordinated behaviours between bodies during common movement. The thinking is that components of human movement systems naturally entrain each other (e.g. bodies walk in step, runners run in step, applauses turn into in-timed claps) as they move to, and create, one common rhythm. It is posited that even if participants are not consciously aware of these processes happening, synchronization can induce feelings of togetherness and happiness, can improve communication and levels of physical performance (Andrews, 2017a).

Momentum

According to the laws of mechanical physics, linear/translational momentum is a motion force that is a combination of an object's mass and its velocity ($p=mv$). Momentum is a vector quality, so for it to exist and be measured, an object must move uninterrupted in one direction for a known time and distance. In terms of practical application, in engineering momentum is most often considered in terms of the energy required to create it (to get objects moving forwards) and to kill it (to stop objects). Momentum, however, equally occurs in the social world, albeit in forms that are less mathematically precise. Four forms in particular, although highly interrelated and often co-existing, are recognized in both academic and public domains: economic momentum, behavioural momentum, circumstantial momentum and felt momentum.

The first, economic momentum, is an established idea referring to momentums in increases and decreases in variables such as market share, overall growth, values, turnovers and profits, all of which are highly quantifiable (Chordia and Shivakumar, 2002; Cooper et al., 2004). In human and economic geography it has long been recognized how these variables might conspire to create physical development and spatial momentums on the ground (see Colby, 1933, for example, on centrifugal and centripetal forces in processes of urbanization). The second form, behavioural momentum, describes the tendency for human behaviour to persist due to its own physical and mental impulses (for example, in career decisions and trajectories), and/or in how one human activity potentially leads onto another with implications (such as for income). The third form, circumstantial momentum, is where, due to the imposition of structures or changing personal or social circumstances (both often beyond an individual's full control), one change for them can cascade into further changes, sometimes in positive directions but oftentimes in negative ones. The fourth form, felt momentum, transpires through, and is experienced as, the basic physicality and force of bodies and objects moving in particular directions. One can experience, for example (as suggested in the Preface) the action and feeling of being swept up, of moving with and adding to the prevailing momentum (and thus the sheer joy of momentum itself, or the discomfort or fright of momentum if the direction is unwanted). Alternatively, one can experience the action and feeling of inertia (there being a lack of momentum), or of moving and working against the prevailing momentum. These positive and negative experiences are often in relation to what is going on immediately, although they can also be about longer-term events. For example, people experiencing consecutive failures or absences might describe feelings of being stuck or in a mire.

Notably, an important category of behavioural and felt momentums – as conveyed in psychology and occasionally in sports and fitness geographies – is performance momentum. Closely aligned with the aforementioned idea of flow state, it is more about sustaining achievement (in lay terms known as being 'on a roll') (Andrews, 2017a). Performance momentum is hence a dynamic intensity characterized by an enhanced state of motion, grace and success. It involves the co-existence and synchronization of mental subsystems (such as excitement and confidence), but also beyond this the positive recycling of emotional and physical energy; the force and intensity of performance momentum propelling individuals and groups towards their goal, allowing them to perform as more than the sum of their parts (Alder and Alder, 1978). Meanwhile, one residual outcome of performance momentum can be a positive psychological mindset; confidence and affective memory that remain in individuals or groups after the end of a specific event until the start of the next. This facilitates the reintroduction of performance momentum once a new event commences, allowing 'the hot hand' to continue potentially over multiple events (Andrews, 2017a).

Momentums in health geographies

Economic momentums arise in traditional health geography in considerations of the rate and forcefulness at which policies for health are developed, and/or markets for health are developed and/or at which services for health expand or contract; this, for example, being a common concern of studies of services for older people (Joseph and Chalmers, 1996; Andrews and Phillips,

2002; Skinner and Joseph, 2011; Hanlon et al., 2014). Otherwise, at more of a macro scale, economic momentum has surfaced in research on the developing world as a quality of the development process and its impacts on health systems and population health (Phillips, 1990; Phillips and Verhasselt, 1994). Meanwhile, across health geography, behavioural momentums are often articulated whereby, for example, one human activity (such as drinking alcohol or taking drugs) leads onto another (such as casual sex) with health implications (such as hepatitis or HIV transmission) (Luginaah, 2008). Finally, circumstantial momentums – progressive and often felt – arise for example as momentums that build as older people move, or are moved, spatially between care settings and services of increasing comprehensiveness in line with their declining physical and mental health status. Although most studies focus primarily on the transitions themselves and their consequences (Magilvy and Congdon, 2000; Lee et al., 2013), certain studies do illustrate how older people negotiate and deal with such challenges – including the momentum of their own declining physical and mental health – either on their own or with the aid of good design and intervention (Kong et al., 1996; Gant, 1997; Rowles, 2000; Wiles et al., 2009). The circumstantial and felt momentums in the daily lives of people with chronic illnesses are also a popular concern (Moss, 1997; Moss and Dyck, 2003), these studies conveying how, in many cases, people's lives might be more appropriately characterized as being stuck or immobile (lacking in momentum) or indeed shrinking (possessing momentum in an unwanted and reverse direction) (Dyck, 1995; Pain, 2001; Crooks, 2007).

More-than-representational health geographies deal with the same kinds momentums but seek to animate their necessary physical and immediate dimensions and the forces produced. One area of research here is focused on verbal and physical momentums in care practices. For example, in Andrews and Shaw's (2010) study of the visualization practices nurses and doctors use when dealing with clients with needle phobia, the respondents talk about constructing conversational momentums that overlay and encourage greater physical momentum in clinical procedures. Indeed, respondents in this study aimed to take client's minds 'elsewhere' (out of the immediate clinical environment and context), allowing themselves to more quickly proceed with and complete the injection. In particular, respondents articulated their use of quick rapid-fire questions and small talk that related to work, leisure or popular events and almost always places. One comments that her speed of delivery helps create a quick distraction and procedural momentum:

> If I need to get through a few of them, there's no time to sit each one down and explore their past. I might ask a child where they live, where they go to school, whether they like to play in the park outside here. They are distracted and I suppose will think about those places. It's all very quick. It works better that way.
>
> (2010:xx)

Elsewhere momentums in public space and everyday life are a concern in more-than-representational health geographies, a popular subject here being movement activities. In Middleton's (2010) aforementioned study of the embodied experience of urban walking, for example, one of her respondents writes in his diary about his specific experiences and the tactics he uses to maintain his momentum:

> This morning I feel tired almost as soon as leaving the house, which doesn't bode well but I know there's only twenty minutes before I get to stop, so I persevere. And that's what a lot of urban walking is all about; persever-ance and momentum. Momentum is vital. When crossing the road I try to do so without having to stop, without having to turn sharply, without having to break my stride, in the morning I'm on autopilot, I can go for ages without even realising I'm walking sometimes, I just point myself and go. My legs know which route they're taking without too much input from me. Another part of the momentum is dodging without effort. Picking the racing line around corners, gliding round the edge of puddles and swing-ing out of the wake of the travelling cigarette. It's an annoyance.
>
> (2010:583)

Similarly, Wylie (2005) talks about the body practices necessary to maintain momentum in long-distance coastal walking, particularly once negotiating terrain becomes physically challenging. His research describes how, in such moments, the greater purposive effort and sound of the body (and attached objects) become part of an overall closer relationship between body, self and landscape:

> Walkers on the Path very often find themselves in such a close visual, tactile and sonorous relation with the earth, the ground, mud, stinging vegetation. Topos in quickstep, in a succession of short sharp bursts. The sound of breathing and the rustle of the rucksack shifting about awkwardly no longer emanate from inside; the affirmation, as it were, of intentional action and effort. Instead they become an anonymous soundtrack through which movement is realized. As if the pre-established boundary between self and landscape, subject and object, could become soluble, osmotic, in the engaged, involved practice of walking.
>
> (2005:239)

Echoing these points but in the context of a different activity, Lorimer (2012) describes his encounters with troublesome and challenging topography in long-distance running, his regular rhythm being forced out by his effort to maintain momentum:

> As a lifelong long-distance runner my encounters with topography are forever a bareknuckled affair. Changes of surface and slope are met with

short jabs and clenched fists – left, right, left, right; as if tattooed with 'LOVE' and 'HATE'. And where swinging fists lead, pounding feet immediately follow. Earthly landings: occurring in countless thousands.

(2012:83)

Imminence

Imminence is important not only as a state and apprehension experience in itself, but because it helps scholars think about the different directions and trajectories of bodies and objects that proceed events. Imminence can be an outcome of other qualities of movement-space (such as rhythm, momentum) and it often arises because of their energy and forwards movement; because they propel and give some notice. In terms of timing, imminence refers to the close future arrival of something. In terms of likelihood, that arrival is certain/impending, or at least it is thought to be. What exactly might be imminent in life varies greatly, from the initially non-physical (such as the pre-dated arrival of laws, rules or policies) to the physical (bodies and objects). Imminence is particularly important in terms of how the arrival is expected or unexpected, how bodies and objects are prepared or ill-prepared for it, and the actions they take or do not take prior to, and on, arrival.

Imminence is important but not special in NRT, which thinks of the world as being in a state of permanent imminence; everywhere in it being forever incomplete, and everywhere in it being open to something new, that will almost always arrive. This goes back to the way in which NRT conceives and understands the body. As Greenhough (2010) suggests (and as noted in Chapter 2),

Figure 7.2 Imminence

rather than thinking about the body as something that 'is' (with particular qualities), NRT thinks of the body as something that 'does' (with agency) that 'might do' (possesses potential). This is a body open and waiting for something to happen and open and waiting to react in some way. Indeed, as Cadman (2009) argues, NRT strives to think in the same configuration as life; as a series of infinite 'ands' which add to the state of affairs. Thus, in addition to being concerned with what is happening, NRT is just as much concerned with what might be expected next, and after that, and after that. Moreover, NRT is also concerned with the jumps forward themselves, the spacing that enables and introduces the next moment. These ideas are reflected, for example, in scholarly debates on apprehension, anticipation, pre-emption, preparedness (Anderson, 2007, 2010; Merriman, 2013) and on the momentary, often microsecond, bare/still points between action and performance (Thrift, 1997, 2000). As Boyd (2017) suggests, for Whitehead and other processual thinkers, the ongoing creation of 'the new' is made possible by 'prehension'. This is a feeling which, although shaped from knowledge of the past, looks forward to potential futures. Importantly, as life moves forward, although there is always something new to be prehended in the world, there is also always something lost at the same time, because no person (or even technology) can fully prehend their past.

Imminent health geographies

Traditional health geography recognizes many things that are imminent in health and health care. Most research on disease and service distributions, for example, is motivated by a realization (whether stated or understated) that, unless attention is paid by officialdom, a situation of some description is impending, whether this be an under-provision of services, an epidemic, or the neglect of a particular client group. Meanwhile, at a micro scale it is realized that many imminent things are important to individuals, whether this be ageing, birth, death, a medical test result, a new regulation, and so on.

A particular interest in more-than-representational health geographies has been the imminence of landscape (and other bodies and objects that constitute it) during movement activities. Cook and Edensor (2017), for example, describe the practice of night cycling and participants apprehending, partially and intermittently, landscape emerging at them from the dark: participants using their senses, adjusting their expectations and preparing their muscles for what is ahead in a focused effort to maintain their safety and progress. Here, as the authors suggest, the landscape is always in formation and motion, it feeling for participants as though landscape is emerging at them (rather than them moving into, and past, landscape). The authors' field notes read:

> I don't know what's out there beyond the short beam, whether behind or in front, from side to side. Out there is full of uncertainty, but my limited range of vision means I am constantly intrigued by what's beyond the

beam: what's out there? It seems mysterious and unknown. I don't want
to know how the landscape looks, I don't want to map the terrain.

(2015:9)

Similarly in Simpson's (2017) study of the affective experience of commuter
cycling, respondents anticipate the environment; notably, the movement of
cars and pedestrians which, if not observed and countered, would result in
collision and likely injury. Simpson suggests that the cyclists' response tactics
range from quite conscious apprehending, assessment and course adjustment
to 'emergency weaving' as an almost automatic bodily reaction. Notably,
although cycling is quite popular as a form of active transportation, research
shows these kinds of processes occurring across a range of more specialist
sports and fitness activities: in the different imminences, for example, of the
hit coming and the martial arts fighter bracing (Green, 2011), the next automatic move of the climber (Barratt, 2011, 2012), and the forthcoming obstacles to be navigated by the runner (Lorimer, 2012; Cook et al., 2016). These
studies convey the sense of apprehension, human reactions and consequences
as movement-spaces unfold.

Other research has a quite different engagement with, and articulation
of, imminence – instead being focused on how potential physical futures
are affectively anticipated. In the context of her aforementioned critique of
anti-obesity policy and associated media imagery, Evans (2010), for example, describes how public health messages are conveyed and adhered to, not
through the presentation of dry statistical facts about what the future likely
holds if one is fat, but instead through the poor health states that often result
from being fat being conveyed in images and messages that animate the affective experience of these potential futures. This allows individuals to anticipate the basic feeling of a possible (virtual) 'unhealthy' future. In another
study Andrews and Chen (2006) considered bullying as a spatial practice.
The authors observe that, anticipating the physical feeling of being bullied,
victims of bullying often employ spatial tactics in order to avoid bullies and
the 'tyrannical spaces' they create and occupy. Indeed, tyrannical spaces
affectively and virtually ooze fear; this fear often being played out in the vic-
tim's mind as a possible encounter and what might happen. Taking the idea
of possible encounter further, Andrews (2011) developed the idea of needle-
phobic encounter space. Respondents in this study apprehended, both physically and emotionally, what might happen to them in medical spaces, one
stating:

I feel nervous when going into a hospital, in a panic and frightened.
I can't get my mind off it. I can't talk about anything because it dominates
my mind. Physically I sweat, my heart pounds, I tremble. Most of the time
I cry. I can't stop myself crying. I have to apologise to the nurse, let them
know that its not them.

(2011:881)

Meanwhile in the same study by Andrews, respondents claimed that they made adjustments and employed tactics in order to improve their experiences, ranging from total avoidance to undertaking complex psycho-spatial 'dry runs', one commenting:

> When a doctor tells me I will need to get my blood work done, I immediately get anticipatory anxiety and start negotiating in my mind if she is right to ask for certain blood work. I don't usually outwardly challenge it though. Then I think to myself, where should I get this done? Will they use a big needle or will they listen to me and use a butterfly needle; the smallest one usually used on children? Then I take control of the situation, in some pathetic way, and decide on where I will get my blood drawn. When I get there, I sit in the waiting room rehearsing how nicely I will ask the nurse to use the smaller needle and make sure I tell her that I have a needle phobia. Then I get in the chair and I make good eye contact with the nurse and say 'I have a big needle phobia. Is it okay if you use a butterfly needle?'
>
> (2011:881)

These latter comments bring us nicely to the next quality of movement-spaces, encounter. This is the moment when what was imminent actually arrives, when realignments happen and new futures emerge.

Encounter

In science an encounter is understood as a collision between two or more objects, often of interest being the 'impact': the resulting force and the way in

Figure 7.3 Encounter

which energy is expelled (through, for example, heat, light, sound and movement). Not too far from this understanding is what Deleuze, and thus NRT, specifically conceptualizes as 'events'; meetings of bodies and objects moving on specific planes with specific trajectories. These being physical, pre-personal, intensive singularities and important spatial, temporal and processual moments (Duff, 2014a). In terms of their duration, encounters (events) can vary in length according to the nature of the things encountering each other, ranging from the almost instantaneous to longer timeframes. In terms of the forces involved, these might be considerable or negligible (bodies and objects possessing different masses, speeds, rhythms and momentums determining the energy expelled). In some cases encounters might be predictable, particularly if the bodies and objects involved have been formerly enrolled and motivated in known directions by existing structures (such as policies). In other cases, however, encounters might be random, the bodies and objects involved being motivated more by their own qualities, purposes or impulses, setting themselves on collision courses.

The outcomes of encounters are important because they provide a moment/stage for the transformation and reinvention of bodies and objects in relations of becoming (Duff, 2014a). Indeed, new realities are produced by encounters; new movements, directions configurations, and potentially even new bodies and objects. Each encounter in life is significant because each is unique (for example in its physical and sensory character) and because each is never contained or isolated, being part of a greater web of events (i.e. each is made by previous encounters and influences future encounters) (Duff, 2014a). For Deleuze, the existence of constant and unique encounters (events) means that life is made through 'discontinuous becoming'; a complex and lumpy process, rather than a clear and linear one, involving progress and retreat, changes in direction, increases and decreases in capacity with no final mature state – the energy involved never disappearing, it merely changing direction and/or form (Duff, 2014a). In short, then, encounters occur whenever wherever there is matter and life, making the world and creating the future.

In research terms, the idea of events/encounters, understood in these ways, is important because it creates an ontological primacy of events over essences or substances, and it creates a vision of the world as composed of happenings rather than of things. Moreover, it creates a realization that on a micro scale, things do not 'happen to' individuals, rather they are part of intensive happenings (Duff, 2014a). As Duff suggests, the individual does not gain a full sense or perspective on any event, rather they only sense parts of it in terms of its relations to their body (so conscious sense does not exist outside the event but is foremost in the world itself). Importantly, as Duff notes, two factors embed encounters (events) in the human life course in particular. First, although encounters are physical and affective, they do intermingle with other forces, relations and certain fragmented subjectivities. So, for example, as Duff suggests, a postnatal bonding event between mother and child involves their bodies in close proximity – hence touch and affect – but also conscious feelings and emotions (such as affection, trust, anxiety, satisfaction). Second, although as noted each event is unique – ensuring new fragments of affective memory and subjectivity are created – repetitions of similar events act as moments of learning. Indeed, through repetition, the intensive properties of affect become familiar, whilst at the same time related subjectivities become more clearly thought out and known. This being, as Duff suggests, a basis for simultaneous biological and cultural human development.

These lines of thinking are carried forward more extensively into considerations of social contexts by Dawney (2013), who considers some encounters as 'interruptions'. She argues that interruptions – which are momentary, and almost always to some extent physical – disrupt the flow of affective less-than-fully conscious experience, as humans suddenly become more consciously aware of their actions and surroundings. Yet, she notes that interruptions provide an opportunity for humans to take stock, to focus consciously on what went previously and what might be next; to interplay the affective with the cognitive and subjective. Interruptions, for example, might reduce the 'naturalness' of particular performances, but also facilitate bodies and minds moving in new or previously unexpected ways (Dawney, 2013).

Encounters in health geographies

Diverse forms of encounter are mentioned across many fields of health geography, although they are not often the main topic of interest in themselves; various encounters being known and shown to be happening yet not explicitly conceptualized or unpacked. These encounters might be, for example, between businesses and a new policy, mode of regulation or financial arrangement (Corden, 1992; Andrews and Phillips, 2002; Ford and Smith, 2008), or between different sectors, specialities or paradigms of health care (Clark et al., 2004; Andrews and Shaw, 2012), or between settings for public health care and incoming corporate principles and initiatives (Kearns and Barnett, 1999, 2000). Moreover, they might be between individuals or populations and such things as infectious diseases (Shannon and Pyle, 1989; Affonso et al., 2004), health care settings and services (Taylor et al., 1979; Curtis et al., 2007), health professionals (Liaschenko, 1997; Andrews and Evans, 2008), restorative, therapeutic and healing places (Smyth, 2005), and substances, objects, contexts and environments which are either harmful or feared (Percy-Smith and Matthews, 2001; Wakefield et al., 2001; Davidson, 2000, 2005). In each case, the downstream outcomes of encounters are considered for those systems and people implicated in them.

Emerging more-than-representational health geographies, in contrast, display a more direct interest in encounters, particularly in terms of the immediate physicality and feelings involved. Indeed, here there is a fundamental realization that the prime mechanism for bodies becoming healthy is the encounter (event), its experience and outcome. Encounters arise in three broad interrelated fields of study: health care and health interventions, everyday contexts and life, and movement activities. With regard to the first, health care and health interventions, Andrews et al.'s (2013) study of CAM practice uncovered, for example, the importance of initial encounters between therapists and their clients, and the tactics the former employed so that their bodies are immediately experienced and read in the right way. Therapists were aware that their bodily expressions, gestures and movements and positions can subtly communicate a sense of open-ended time, trust, understanding and healing at this critical moment. One comments:

> Eye contact, smile, a handshake, all those things are really, really important. I think they just set a tone, allow an opening to give and take information and start to build trust. That's part of creating a safe environment for the patient... you're almost creating an environment where you invite the healing impulse in a way. Where you cultivate a healing impulse.
>
> (2013:106)

In forms of care, the importance of encounter goes beyond first impressions, however, equally relating to the nature of treatment or therapy itself. In this regard Paterson's (2005) research on therapeutic touch describes the taking place of a powerful physical inter-body encounter and the emotional release it evokes:

> My first Reiki massage, ever. Anxious because of deadlines, a hundred things whizzing through my head... Then the massage begins, a curious mixture of touch and non-touch. As this is going on, something strange and unexpected starts to occur, then starts to surge uncontrollably; a welling up that becomes an outpouring. Along with a feeling of incredible release, I start to cry and find I cannot stop.
>
> (2005:161)

On a similar note, Atkinson and Rubidge (2013) describe an arts-based intervention for wellbeing: specifically, a mask-making activity for school children that emphasizes skills, creativity and expression. Although central encounters were guided by staff and smaller encounters were improvised by children, each contributed in their own way, together making the overall wellbeing-inducing practice. The authors reflect:

> Even the most specified of tasks and activities required a certain spontaneity and creativity of movement. Thus, in making the masks and drawing the expressions, each child moulded the play-dough and manipulated the pens to their own designs [the choreographer involved in the research project, commented] 'The children are making decisions about the expression of their mask – happy, overjoyed, angry, sad, confused, serious, and so on – while their hands are busy with the manipulation of

the play-dough. Hands rise, fall, pat, poke and rub, fingers press, prick and hold. The play-dough is rolled, shaped, pushed, pressed, scrapped, placed, lifted, squashed, flattened, and plumped up. ... some, seemingly almost unconsciously, doodle with left-over play-dough, making for incidental finger stories. One child is singing a made-up accompanying song that includes the words squishy, squishy, this is squishy.

(2013:9–10)

In the above research the creation of new therapeutic possibilities is the main outcome of encounters. New possibilities can be very different however, such as enhanced participation in research processes and change making. Indeed, as described previously, Kraftl and Horton (2007) show how they facilitated 'the health event', a workshop/meeting where the findings of a study on young people's health needs were shared with those who had participated – the encounter producing formal feedback but also its own new life, ranging from energy and enthusiasm to promises for future events and actions.

Whilst the research noted thus far provides examples of expected encounters and intended outcomes of health interventions, other studies show chance, organic and unexpected encounters and outcomes. Justesten et al. (2014), for example, undertook a performative observation of two hospital wards, and describe a moment that broke the silence, regularity and formality of the setting, moving life there in new directions:

One of the most surprising events took place one afternoon just before a health care professional was about to finish her day, as she was off to attend a date that she was excited about. I was sitting at my table reading when she walks out of her office. Suddenly and as a surprise, she starts singing, dancing, and spinning down the ward corridor. In that moment the atmosphere changed into one of joyful energy. Her embodied light movement and her way of filling out the ward corridor changed the ward's 'soundscape' by downplaying the buzzing sound from the freezer and the venting system as well as downplaying the institutional light-blue stripes at the ward walls. Instead, the ward 'soundscape' became filled with human activity as healthcare professionals were laughing and their voices rose cheerfully while they went into patients' rooms with lighter and faster steps, asking them if they would like coffee or if they needed any help. Furthermore, patients not condemned to beds popped out from their rooms and attended the coffee-trolley or the sitting area whilst chattering.

(2014:109)

In the second engagement with encounters (everyday contexts and life), it is recognized that some of these are at least initially purposive. Duff (2014a), for example, explores spaces of drug taking. One of his respondents talks about taking ecstasy and it leading to enhanced conversations and socialization:

> [T]he part of the drug that is really appealing is the idea that you have random connections, that's fun you know. I guess you, um feel able to go and talk to people at random. Like you'll be high and you'll go out onto the street for a cigarette and it'll be raining crazy hard and you'll be talking to this random person waiting for a bus or whatever, and that'll be the best part of the night.
>
> (2014:138)

Other encounters reported in everyday life are however more accidental and organic. In an aforementioned study Andrews et al. (2014), for example, consider experiences and moments of wellbeing. The following field note from their study illustrates encounter and new realities created thereafter. It is one of the researchers' baby cousin's birthday party at a local 'jungle gym'. She walks into the indoor playing area where his party has already started and stops abruptly at witnessing the chaos that surrounds her:

> **Crash, bump, whizzz, bang.** Everywhere I look a child running or climbing or crawling. The area is filled with shrieks of laughter some reaching higher than the rest piercing my ears with their high notes. Its action, movement and young energy everywhere. I soon find myself the target of the kids in the ball pit throwing the multicoloured balls at me. As they flash by I try to dodge them one at a time, this way that way, left, right, right left. I laugh as a plastic ball ricochets off my head **'booooong'.** Without thinking I join in the throw fest and start to return the balls which are light and easy to the touch. Left hand pick up, right hand launch, left hand pick up, right hand launch – over and over. I'm smiling, throwing, missing and hitting and feeling young again.
>
> (2014:215)

In the third engagement with encounters (movement activities), it is recognized that repetition of encounter is often part and parcel of the overall experience. Stephens et al. (2015), for example, consider disabilities in children and how their bodies learn through encounter, in some cases involving trial and error and even physical pain. The authors recall:

> As we were walking [child] fell many times. He literally fell hard to the ground, once on the pavement and a few times while we were walking through the field to get to the park. After he fell, his sister instructed me not to help him get up – because he had to learn.
>
> (2015:202)

Elsewhere in Cook et al.'s (2016) study of urban jogging, one participant talks about encounters on pavements and roads and overcoming obstacles, whilst another talks about their tactics:

> I don't really like it when it comes to running on the pavements... and sharing it with other people because you have to get out of their way and they have to get out of your way and they don't see you and you end up crossing the road rather too frequently. It is quite dangerous because I think drivers expect runners to get out the way, not to be there.
>
> (2015:20)

> I kind of like duck to one side as an indication saying 'I'm leaving you space to get past this side – kind of take the hint or I will run into you!
>
> (2015:21)

Similarly, in Simpson's (2017) study of the affective experience of commuter cycling, what he terms 'proximity affects' occurred. Cars, for example getting too close, causing annoyance and responsive gestures (hostile glances between drivers and cyclists); cyclists finding spaces between cars and between pedestrians and weaving, and cyclists using their bells to give notice and to avoid collisions. Taking thinking about encounter even further, Wylie (2005) meanwhile talks about encounters with the vast/far vs close up/closed whilst recreational walking. He notes, for example, what happens when he stops moving forward on clifftops (perhaps to rest); that he suddenly becomes conscious of the vast vista to the horizon. He reflects that on one occasion, whilst walking in a woodland he stopped and suddenly became conscious of, and attentive to, details and textures of small and specific things such as individual branches and trees. Wylie also reflects on stopping and seeing a wild valley:

> The quotidian rhythm of walking, connoting an understanding of landscape as a milieu of corporeal immersion, is counterposed by a visionary moment of drama and transfiguration.
>
> (2005:242)

These studies show us that in all movement activities there is encounter, something colliding, something changing, something new emerging. Wylie's points here about stopping and being stationary, and the contrasts it produces, lead nicely into the final quality of movement-space considered in this chapter.

Stillness

According to physics, for an object to be still/at rest, it must lack any measurable or perceptible speed. Newton's First Law of Motion states that an object will remain at rest unless acted upon by an external force. However, in practice nothing in life is ever completely still – at absolute rest. An object will vibrate and move with subtle atmospheric or environmental effects, its electrons and atoms always move and ultimately it is situated on the earth, which is rotating around the sun. As a result, stillness is really a human perception relational to dominant/majority objects and relative to the observer. Moreover, stillness is not just a measurable state, for human bodies it is also a sensation and experience felt physically and mentally, these facets making it of particular interest to emerging traditions in human geography, most notably to the new mobilities paradigm and to NRT in particular (Bissell, 2011; Bissell and Fuller, 2011). With regard to NRT, stillness is important because, as shown in the previous chapters, so much of the approach is focused on energies and aspects of movement (rhythm, speed, momentum etc.). Stillness then is the counterpoint, the reference or the contrast to these. It is certainly not like them, but has to exist for them to exist.

Still health geographies

Stillness is traceable in across many fields of health geography. As suggested in chapter 1, the general critique of representational research is that it provides false (still) snapshots in time (for example, of a disease or service trend), even if they are not themselves still. In this respect, then, all health geographies are still health geographies but because of how they fundamentally report the empirical world. More-than-representational health geographies are, however, beginning to address this situation through two interrelated fields of research that show the stillness of situations and moments themselves.

Figure 7.4 Stillness

The first of these fields is focused on coping with, and living, everyday life. Andrews (2018), for example, examines the affective and meaningful experiences of sufferers of Chiari Malformation (as suggested earlier, a structural brain defect most prevalent in children and young adults, with symptoms such as dizziness, headaches, muscle weakness and disorientation). Respondents in his study employed stillness/non-movement as a basic coping strategy to reduce their symptoms, although this equally led to further isolation and removal from daily affective homelife. One comments:

> Sometimes the only way to cope with the pain and disorientation is to stay completely still. Not moving at all – typically sitting. That's not going to make you feel better or socialise or anything. Its just so you can exist with some small level of comfort.

Elsewhere, in terms of everyday life, Bissell (2008) considers the reaching of comfort through 'sedentary affects': bodies touching and otherwise sensing proximal objects (such as chairs) and undertaking subtle performances (such as adjustments) to maximize affective potential. Here, then, the body as a whole might be physically still, yet still moves a little affectually.

Movement activities are an important consideration in this field of research on everyday life and here Edensor and Richards (2007) convey moments of stillness that are important to snowboarders. Even though their sport is known to be fast and adventurous, in stillness lies a momentary opportunity for rejuvenation, a respondent stating:

> They [skiers] can't sit down. How annoying is that? They've got to stand up all day, legs burning, absolutely bollocked. We can just drift along, have a little sit down, have a chill out, have a smoke, have a picnic.
>
> (2007:108)

Meanwhile, studies have shown that stillness of mind can be important to movement activities; such as in the natural, rhythmic spiritual experience of running, walking and cycling. Here the mind might be still, yet the body moving (Bale, 2004; Lorimer, 2012; Brown, 2017).

The second field of research is stillness as a therapeutic practice. Andrews' (2004) aforementioned study of visualization in CAM discusses, for example, how clients are often sitting still, on a chair or table, which is important to practice. A therapist comments:

> I use imagination to relax clients. You can't massage muscles that are tight, so you need some way of relaxing them. Taking a person's mind away is one way of doing it. Because they are lying on their front, lying on a beach is an obvious comparison. I don't go on about it, just make a few suggestions.
>
> (2004:314)

Similarly, Andrews et al.'s (2013) study of CAM shows a field diary entry – involving the centrality of stillness, both physically and in terms of the sound-scape it offers:

> Ann [the therapist] told me that by far the best way to know therapies was to have a go. Choosing reflexology from the three she had on offer, I entered her room and sank into her large leather reclining treatment chair. This felt like being held in a giant glove; it gently supported and cradled every tired limb. We talked about reflexology whilst I was having the treatment, although I was becoming more and more relaxed and less and less bothered about asking questions (which suddenly seemed forced out of place and were getting slower and slower). Before long I dispensed with the back and forth of questions as I could not really do anything with the information coming back. I was totally chilled out by the therapy and background music, my arms heavy, legs heavy, sounds washing in and out like waves on a beach. At some stage I drifted off happily into a light sleep, awakening on my own some time later. Lots of people do that Ann said as she walked brightly into the room, its perfectly okay, although she advised me to stay reclined and sitting for ten minutes to 'come back'.
>
> (2013:106)

Elsewhere, in Conradson's (2007) aforementioned study of therapeutic/spiritual retreats (abbeys), a core quality of the setting and overall residential experience was stillness. As Conradson describes, the reaching of an embodied state of calm and inner tranquillity in residents was assisted by the state of calm in the other human and non-human elements present (such as the quiet, echoey surrounds of a 'shrine room', comfortable cushions for sitting, and blankets for additional warmth). Conradson argues that stillness allows for new modes of feeling yet, feeding into the conscious realm, provides opportunities for clarity of thought. Conradson claims that paying for these types of places/services is part of the commercialization of affect, and stillness in particular, in western societies; part of the commodification of modes of feeling. He suggests that whilst in society's main domains (such as workplaces), stillness is considered unproductive, stillness in 'elsewhere' places (such as retreats) is considered productive, is more easily achieved and has a market exchange value. Meanwhile, Conradson's (2011) paper on the same retreats describes the more personal processes at play. How, to the individual, stillness involves distancing 'there and then' (their busy life) by focusing on the 'here and now' through centring the mind on the body and its current environment. Thus he suggests that this helps break cycles of mental rumination and helps open up more lucid states of consciousness. Indeed, overall then stillness can be obtained through (i) removing oneself from a non-still environment, (ii) placing oneself in the new still environment, and (iii) adjusting one's mind to be predisposed to the stillness on offer. The retreats Conradson studied provided structured opportunity for all three of these things.

8   Research practices and future directions

As Thrift (2000:222) argues, 'in non-representational theory what counts as knowledge must take on a radically different sense. It becomes something tentative, something which no longer exhibits an epistemological bias but is a practice and is a part of practice'. It involves attending to differences in ways that encourage a re-presencing of the world we 'know' rather than the world we 'think about' (Dewsbury, 2003). It involves opening up the field of geographical inquiry to the possibilities of performance – the arts of experiment and the art of written communication (Thrift and Dewsbury, 2000). It means challenging the 'know and tell' of much of human geography and embracing failure, because we can only ever produce partial understandings of a world that simply takes place (Dewsbury, 2009).

(Boyd, 2017:51)

Methodological priorities in NRT

It has been noted that, beyond the increased deployment of qualitative methodologies and related innovations, the cultural turn in human geography in the late 1980s and early 1990s had very little impact on how geographers actually did empirical research (Latham, 2003; Patchett, 2010). But, as Thrift (2008) argues, and as suggested in Chapter 1, in contrast NRT actively attempts to escape what it regards as the excessiveness of traditional social science approaches and their mission to predict and/or understand everything they encounter. This escape, as Roe and Greenhough (2014) observe, has necessitated changes to methodologies that are further encouraged and refined by two factors. First, by the nature of the information sought; NRT being less concerned with what people say/represent (captured by talk-based methods) and more with what they do in living their lives (the engagements and senses involved). Hence, under NRT methodologies, no longer do researchers have to capture reason; instead they have to capture action. Second, by the fundamental understanding that certain knowledge – whether a subject's or a researcher's – is gained processually through lived embodied experiences, often less-than-fully consciously: through senses (seeing, hearing, touching, tasting, smelling); and, more broadly, through doing and moving with (i.e. not only through reading texts and listening to words).

As a result, NRT methodologists can be radically experimental, adopting a number of mindsets and practices. With regard to mindsets, they are prepared to provide open-ended accounts, to show and tell but not fully explain (Simpson, 2010). Moreover, they are prepared to be led by events, not led by the intent to repackage them. As Thrift (2008:12) notes, they want 'to see what will happen, to let the event sing to [them]'. With regard to practices, they are prepared to be more daring, conceptual, playful and expressive. This involves both experimenting with new ways of engaging the world, capturing data from the world and ultimately telling the world; their aim being to 'witness', 'act into', 'build', 'change' and 'boost' the world, and help it 'speak back'.

In terms of explanations of these five aims, witnessing is paying close attention to the unfolding of movement-space, including the numerous qualities and events that together create it (Dewsbury, 2003). By doing this, it is hoped that the emerging data and reporting might have a fidelity and faithfulness to what is happening; not re-representing it, but 'presencing' it (i.e. summoning it up and making it present once again). As Latham (2003) argues, with NRT established methods in human geography do not have to be abandoned, they just need to be adjusted and/or augmented and/or combined and made to dance a little more, involving for example go-along-interviews, photography and video. Acting into life, on the other hand, is based on a belief in the power of the active world, and that researchers need to co-experience it with their subjects. Moreover, it denotes a close relationship between the researcher and what is happening in the field, and a blurring between the role of observer and what is observed (Dirksmeier and Helbrecht, 2008). Indeed, with NRT, method is itself a performance that does not so much study a social reality through the acquisition of data, but more 'does' a social reality and lives the data (Vannini, 2015a). For example, in addition to using some of the methods noted above, interviewing can be more about the interpersonal interaction than the topics discussed, and participant observation can be reversed to 'observant participation'. This involves more than watching: doing the same thing as the subject, getting embroiled in their efforts and investments, becoming infected by the energy and spacing of unfolding situations, and actively changing the course of events (Thrift, 2000; Dewsbury, 2009; Boyd, 2017).

For certain commentators, at the heart of much NRT lies a political motivation to rally against and challenge the ways in which peoples' lives are manipulated by institutions under authoritarian neoliberal democracy and advanced capitalism, and particularly the way in which they insert mass-produced aesthetic encounters into our lives. Thrift (2011) argues that new playful and experimental methods are needed to engage this new technologically driven, entertainment-orientated world that probe and provoke awareness in unconventional ways to foster new ways of associating and some degree of understanding. In short, attempting to join, as it were, the new world in its own game; both challenging and celebrating it (Thrift, 2011). Thus it follows that building, changing and boosting the active world, and helping it to speak back, are all about introducing new realities into life, both in the field and through forms of knowledge translation. This is an important and somewhat unique methodological priority of NRT, not always part of every study, but seen as an ethical action in itself (Vannini, 2015a). The agenda towards building, changing, boosting and speaking back has been partly served by some of the methods mentioned previously but also, quite specifically, by the development of arts-based research (often including theatre and dance) and a more general agenda towards an activist and public scholarship. These approaches have obvious connections with the wider move towards 'experimental geographies' in the parent discipline (Last, 2012), whereby the researcher is envisioned, not as a critical bystander as space-time unfolds in front of them, but as an active, creative producer of space-time (Gallagher and Prior, 2014). We see these kinds of experimentation, for example, in the ways that NRT regards the arts as both a topic and inspiration at once (Thrift, 1996; McCormack, 2013; Andrews and Drass, 2016). As Thrift (2008) notes:

> I believe that the performing arts can have as much rigour as any other experimental setup, once it is understood that the laboratory, and all the models that have resulted from it, provide much too narrow a metaphor to be able to capture the richness of the worlds.
>
> (2008:12)

McCormack (2003) argues that recognizing NRT methods as a practice of intervening is important, because otherwise NRT runs the risk of becoming a tradition that produces a reconstructed representationalism of the banal, ordinary and everyday. Moreover, as Greenhough (2011b) explains, such intervening is part of the adoption of an ontological politics which, she suggests, can be thought of as a conviction that a purely material/physical world – or a 'real' world – that precedes human practices and interactions does not exist. Rather these practices and interactions shape the world, which has a future not determined by conditions and always open to re-negotiation. One could argue, however, that there are also more practical motivations behind such practices based on the current political environment. Indeed, in a world where right-wing/nationalistic/neoliberal movements and change have taken hold, there is a clear need not only to come up with new liberal ideas and visions, but to work in new and inspiring ways. In this context, in research terms, if much social constructionism and post-structuralism instructs, 'do this, and do not do that', NRT instead animates what is done and does what can be done. NRT's agenda for change is hence not one based in words and rules, but is one based in action and doing; NRT being an active left.

Methodological challenges and mitigating strategies in NRT

Two methodological challenges have persisted in NRT that have been hotly debated, both concerned with the nature of the active world and the limitations of social scientific methods to convey it. The first is to do with complexity; the subject of study – such as affect – often being an experience produced from, and inclusive of, a wide range of body and object movements and interactions. As Ducey (2007) suggests, full environmental and sensory happenings in life can never be translated directly through language and words – either on the part of subjects or researchers – which tend to simplify and deaden them or go too far down contemplative and interpretative paths (see also Latham, 2003). The second challenge arises because the happenings studied are often less-than-fully consciously acted and experienced; for example in the case of affect existing as a background hum. As Pile (2010) explains, the underlying problem here is one of sequencing, and the inevitable involvement of the subject's and researcher's own cognitive judgements during their conscious observations, reflections and articulations. Indeed, any individual might witness, experience or add to an event but, as soon as they reflect on it, they involve their personally, culturally and historically anchored thoughts and emotions, which evoke a re-created consciousness of it (now one step removed from it in its pure original form (Conradson and Latham, 2007; Ducey, 2007)). In response to both of these general challenges, a partial mitigating strategy involves one or more of three approaches.

The first approach is for researchers to develop an underlying predisposition so that they might be able to pick up what they need to pick up. This involves approaching the world with a sense of 'wonderment' (Thrift, 2008; Vannini, 2009, 2015a). It is critical for them to have a fundamental sense of appreciation and excitement to see, and participate in, the movement and energy in life. This is because they must be able to look at places and situations through eyes unburdened by immediate thoughts about where social structures might be at work and where divisions might be occurring. Instead, they must be willing to take in the foregrounds and backgrounds, feel the energy involved, unsettle and disrupt at the surface, and ultimately look to animate these things to their audience (Vannini, 2015a). As Dewsbury (2003) and Boyd (2017) understand it, the gathering of life knowledge through such wonderment is less about judging or reasoning, and more about apprehending and being open to the 'eventhood of the moment' – the singular, unique and 'flashes of encounter'. Notably, at times, this sense of wonderment is not unlike that found in the popular spirituality movement generally, and mindfulness techniques more specifically, whereby a person might experience brief moments of self-transcendence in observing, touching and listening to their immediate environment. Whereby a person might experience the world, and their place in it, physically in the purest of forms, as free from preconceptions and judgements as possible; free from the constant stream of the mind's never ending self-narrative. Moreover, a researcher might also develop wonderment that extends to great scales. Not unlike the wonderment associated with theoretical physics – or at least its public articulation and popular imagining – this involves appreciating the almost endless physical threads and energies that span and make time and space, and make life as it appears to us, from atoms and molecules to complex bodies and objects, to solar systems and the universe. This all seems like a challenge but, as Thrift (2008) notes, for the practitioner of NRT, the opening and opportunity for wonderment comes partly from the freedom gained by no longer having to fully explain or thoroughly theorize the social subject. Indeed, with these preoccupations and responsibilities taking a temporary backseat, the researcher has the space to be engrossed on a more fundamental and physical level.

The second approach to overcoming the aforementioned methodological challenges is to introduce complementary approaches that provide either secondary sensory insights or new events themselves (such as photography, music and other arts). This might be done so that legitimate attempts are made at relaying, presenting and making realities more than (re)representing them, even though one can only mitigate so far. In sum, as Andrews (2014a) posits, characterizing NRT is its inherent methodological flexibility, experimentation and playfulness and willingness to: (i) produce case-specific hybrid methodologies, so that observations stay focused on the immediate and do not go too far down contemplative and interpretative paths (Patchett, 2010); (ii) embrace methods that move beyond forms of linguistic expression in both data capture and knowledge translation (Patchett, 2010); (iii) 'montage' methods, so they are juxtaposed and inhabit different space-times – not 'triangulating' in a purist/traditional methodological sense, but reflecting the multiplicity and relationality of the world (Latham, 2003). Vannini (2015d), for example, summarizes a number of the above approaches in the context of his particular form of 'mobile' NRT ethnography:

> Fieldwork generally entails travel, and ethnography in the end, perhaps, is little more than a glorified form of travel-writing. Mobile ethnographies have thus begun to make sense of the itinerant and kinesthetic aspects of fieldwork and to portray ethnographic work as happening on the move. Non-representational ethnography is particularly well equipped to handle the kinetic dimensions of fieldwork. Drawing attention to movement is important not only because it evokes the perambulations of ethnographers in all their vicissitudes and complex logistical relations, but also because doing so situates fieldwork in the concrete time-spaces that ethnographers actually inhabit. In fact, all fieldwork unfolds in space and time and as a result ethnographies are positionings, orientations, and *travails* always busy with securing the loss of old habits and coming to terms with being out of place. 'Inhabiting space', like ethnography, 'is both about finding our way and how we come to feel at home'. Like all forms of travel, therefore, non-representational fieldwork engages us – as practitioners and audiences – as a constant re-negotiation of difference and repetition, of the ordinary and unfamiliar. Building from Ingold's ideas on movement, we might then characterize non-representational ethnography as a practice of wayfaring. From this perspective, ethnographic journeys are not planned transitions from the office to the field site but wanderings through which movement speaks. These wanderings are also wonderings which seek out the interweaving storylines binding

self, others, places, and times – lines which, just like ethnographic travel, are dynamic, unpredictable, with no clear roots or obvious boundaries or ends. Through journeying as storytelling, nonrepresentational ethnographers take storylines in flight, out for a walk, along on a paddle, forming 'knot[s] tied' from multiple and interlaced strands of movement and growth.

(2015d:323)

The third approach to help overcome the aforementioned methodological challenges is, if one has to write about happenings, to describe them expressively, including the sensations involved, and be as true and honest as possible to their movement and energy (Cadman, 2009; Vannini, 2015c). As Ingold (2015) describes, in order to do this it is necessary to break partially from traditional academic language and speak the everyday language of the world with the world. Indeed, with NRT researchers tend to avoid locking straight into rigid, exclusionary, categorical academic frames, and instead develop a lively style of communicating with words that creates itself in the image of the very practices it presents (Ingold, 2015). This is not unproblematic because the formal 'realis mood' of writing all academics have been trained in and are accustomed to – which attempts to be authoritative, logical and definitive in its statements – itself justifies the very existence of scholarship, separating it – and thus what academics do – from personal communications, fictional writing and review journalism (Vannini, 2015c). Nevertheless, scholars of NRT suggest that one does not have to accept this status quo, and can also incorporate an 'irrealis mood' in academic writing that can be purposefully uncertain, can create senses of unreality, surreal, possibility and hope, and can help subject matter come to life, reverberating differently with different readers. Moreover, to achieve this, as Vannini (2015c) suggests, one can employ a number of irrealis sub-moods variously including the conditional mood (stating propositions, not certainties, with conditions), potential mood (being tentative and acknowledging other possibilities), fallible mood (acknowledging that some of the world is illogical and cannot be expressed by language, and that the writer can have doubt and ignorance), hypothetical mood (exploring possibilities and posing 'if'), immediate mood (writing as if in the moment, as if things are unfolding right then without a known outcome), admirative mood (expressing wonder, fascination and surprise), subjunctive mood (expressing states of unreality or realities that have not yet occurred) and desirative mood (expressing wants, wishes and preferences). These moods can be read in many empirical papers that employ NRT and in what Lorimer (2007:90) terms the 'quicksilver quality' of Nigel Thrift's explanations and mappings of the field that, he says, rather than getting progressively denser and deeper, are 'forever pushing outwards'. They represent a movement away from a bland professional language that is bare of expression, performance and unpredictability, to a language shared by those who work in the arts – sometimes written, sometimes spoken aloud (Ingold, 2015).

Another way of thinking about new forms of writing is through the analogy of sports broadcasting (as mentioned in briefly in Chapter 1). In this respect McCormack (2013) notes:

> The scenario might be familiar, yet it never ceases to irritate. The screen cuts to the upper torso and face of an athlete: a gold medalist at a major athletics championship; the captain of the victorious team of a fiercely contested football game; the scorer of the winning goal in the same game. Sweating, beaming, animation: joy unconfined, obvious and palpable. And then the question: What does it feel like? Wonderful, absolutely amazing, brilliant. I can't really believe it. And still the sweating, beaming, animation: joy unconfined, obvious and palpable. Changing tack, the interviewer probes and presses for a little more detail, insight, reaction: Talk us through that goal. Forced recollection of movement, patterning, connection. To be honest, I didn't really think too much about it. I think that Gary fed the ball to me and I cut inside and then just hit it as hard as I could... And then a reaching for some sense of significance. Could you tell us what this victory means to you? It means so much, I can't really put it into words. And still the sweating, beaming, animation: joy unconfined, obvious and palpable. So there we have it. Back to the studio for some detailed analysis. The unease generated arises because the scenario is so reminiscent of the ethos, style, and conduct of so much methodological work in the social sciences. At the risk of caricature, a great deal of this work operates through the logics of the postevent interview. [It is] a result of the residual assumption that rigor can be approximated and vagueness reduced only through the construction of a triangulated account of interpretive sense making. The proposition developed [by the author] is that the performative logics of 'running commentary' (play-by-play announcement) on sporting events offer ways of stretching the relation between events and their expression... Clearly the point is not to jettison postevent interviews. Nor to stop talking about events. However, when talking or writing about affective spacetimes of moving bodies, it might be possible to refigure the relation between these spacetimes and forms and styles of talking and writing. It might be possible to move away from, or at the very least supplement, the logics of after-the-event sports reporting.

(2013:117–118)

Further, McCormack notes that conducting play-by-play announcement means being the 'purveyor of actuality'; the conscious creator of word pictures of the unfolding of events. Beyond even the irrealis style, this, according to McCormack, involves additional writing priorities and techniques, including:

(i) Appealing in the first instant to basic movement; its emotion, thrill and excitement.

  (ii)   Understanding and conveying active contexts including background happenings, sights and sounds.
 (iii)   Non-technical description of happenings.
 (iv)   Modulated writing voice which reflects the intensity, speed and volume of the activity; transmitting the rhythms at play.
  (v)   Addressing an audience as if explaining to people present and watching what they are experiencing.
 (vi)   Possessing confidence to use a 'territorial style' which is sympathetic rather than biased.
 (vii)   The use of refrain, including repetition of names or terms or carefully placed stops to create suspension and anticipation.
(viii)   Intervening with analysis in a compatible way that does not kill the action (like the additional 'colour commentator' might do in professional sports broadcasting) (see also Andrews, 2017a).

Methodological innovations in more-than-representational health geographies

Methodological innovations in more-than-representational health geographies have not been particularly radical or leading in comparison to those in NRT-informed human geography as a whole, yet considerable efforts have been made. For the purposes of this chapter, the key studies that demonstrate methodological innovation are organized into five themes which reflect particular techniques and approaches: progressive ethnographies, lively interviews, using new technologies, working with the arts, and styles and forms of writing and presenting.

Progressive ethnographies

Health geographers engaging in NRT have tweaked and adjusted their ethnographic investigations to insure that, during the course of their studies, they are able to capture and convey the immediacies they are primarily interested in. Justesen et al. (2014), for example, conducted a performative participant observation on two hospital wards – cardiology and gynaecology – to live and convey the immediate, embodied, present moments in the hospital meal experience, focusing on agency and events that disrupt everyday practices. The authors' montage of methods includes the combination of extensive photography (over 200 photographs) and twenty-five structured interviews with patients in which they ask them to 'describe what happens' and 'what they (you) do' (rather than what it means).

Similarly, drawing theoretically on performance theory and also empirically on their experiences of working on the UK Medical Research Council's Common Cold Unit, Roe and Greenhough (2014) focus on adjusting their research techniques to produce new knowledges. The authors claim that methodology – both in the science they studied and in NRT – like the very life it engages, is habitual yet also improvised, involving and shaped by sensory and affective capacities. Based on this observation they develop the term 'experimental partnering' as a specific methodological approach that, they claim, is attentive to both habits and improvisation in methods and life; an interpretative mode that is sensitive to and apprehends assemblages and their workings. Indeed, Roe and Greenhough claim that experimental partnering involves a range of intuitive or informally learned practical activity whereby the researcher is always reaching towards a goal, and towards making something work or intelligible. This necessarily involves a process of trying and failing, and trying again another way, even if knowing is never reached. Specifically, the authors explain:

> The 'experimental partnering' approach to studying habits brings the researcher's eye, nose, mouth, hand, body, etc. to be curious in a world that is dynamic, busy, playful as well as awkward, limiting and more than what appears. In this way what the researcher may observe, sense, become entangled with in the field is equally agentive in amplifying a reality that captures the attention of a researcher detecting what is taking place. How, where and when the researcher engages with the 'objects of study' will affect what happens and what could become known.
>
> (2014:54)

With regard to progressive ethnographies, credit should also be given to McCormack's (2003, 2013) work on dance movement therapy (discussed previously in this book, and at greater depth in a later section) and the ways in which it masterfully draws out and animates the refrains, movements and energies in this particular healing practice.

Lively interviews

As suggested earlier, in NRT interviews can be performances in themselves, and focusing on their conduct can produce data as useful as any verbalized message recorded. Although health geography has rarely gone this far off the beaten path with interviews, at times they have been tweaked or variations employed. One such development is adjustments made to help find the affective, Ducey's (2007) aforementioned research on the training of health care workers being a good example. Ducey aimed to find out what physicality and performance make work meaningful for these workers, her interviews being set up to probe the energetic infectious aspects of work life. During conversations Ducey looked particularly for when respondents became animated in giving their responses, mirroring their energetic work encounters: one respondent, for example, exclaiming with passion: 'learning, learning, learning' (2007:198).

In another study Dean (2016) considered the potential of go-along interviews based on her experiences of using them to study a range of lively empirical topics (including adolescent physical activity, the use of green space and older people walking). Dean argues specifically for the montaging of go-along interviews with observation of participants in their field at the same time, which might involve all kinds of mobility (including walking, cycling, public transport and riding in cars). This form of interviewing Dean argues, allows her to consider simultaneously the representational and non-representational (i.e. participants' verbalized perceptions of the changing environment, and their navigation, practices and patterns of interaction through the changing environment). Indeed, Dean suggests that during this process, environmental landmarks might simultaneously trigger both particular verbal interactions and particular physical negotiations or interactions. Similarly, relationships might exist between topics of conversation and ways and intensities of moving. Dean reflects however on the challenges involved, including obtaining ethical clearance with this particular methodology due to the risks to the safety and anonymity of participants, particularly if they are entering the environment purely to participate in the research.

Using new technologies

> There is [increasingly in research] the ability to sense the small spaces of the body through a whole array of new scientific instruments.... Thus, what was formerly invisible or imperceptible becomes constituted as visible and perceptible through a new structure of attention.
>
> (Thrift, 2008:67; quoted in Spinney, 2015)

More-than-representational health geographies have utilized new technologies to capture and animate the movement and energy in the activities and situations they are interested in. Simpson (2017), for example, considers the affective experience of the active transportation method of cycling to work. In his study, twenty-four commuters recorded their journeys on head-cams which facilitated later interview discussion whereby participants talked through the playback with the researchers. In another study that more broadly discusses the potential of technology in researching movement activities and the embodied experiences they involve, Spinney (2015) considers the combination of bio-sensing technologies (such as GPS-enabled EEG sensors) and mobile qualitative methods (such as go-along interviews and mobile video ethnography) to capture the meaningful, sensory and affective. The two approaches, he argues, can complement one another. Whilst the former, he posits, might provide bodily data on intensity/quantity in a quantitative form that avoids a high degree of subjective interpretation, the latter he posits might provide critical insights into the 'feel' and 'quality' of the overall experience (thus both the *what* and *how* of the performance and event get addressed).

A third example of the innovative use of technology is Bell et al.'s (2015) study that uses GPS and geo-narratives to understand everyday routine and habitual green space encounters. Indeed, the authors produced personalized maps using participant accelerometers (to measure physical activity) and GPS data (to measure space covered), alongside in-depth go-along interviews. Bell et al. argue that the montaged methods offered opportunities to build a more nuanced, contextualized and layered appreciation of participants' green space experiences and the wellbeing they evoked. The authors claim that on one level, participants' conscious interpretations of their own practices were accounted for, whilst on another level, visual aids and data complemented this narrative, speaking to their practices that occurred physically and less-than-fully consciously.

Working with the arts

More-than-representational health geographies have engaged variously with the arts with different intentions and for different ends. Richmond (2016), for example, describes a community-based film project developed with Canadian indigenous communities. *The Gift from Elders* was produced to document and preserve indigenous knowledge on health typically passed down orally. The film supported this knowledge through its documentary style that explored youth journeys as they learned from elders over a summer period. Richmond reflects that this is a 'decolonising methodology' which puts indigenous communities and their concerns, aspirations, approaches and leadership at the forefront of her research. A materialist interpretation being that the film is an object that circulates knowledge and shares information within and beyond the community (on beliefs, traditions, previous injustices and misinterpretations), but the creation of this object itself leads to self-reflection on the part of those researchers and indigenous people involved. *The Gift from Elders*, as the author suggests, creates its own space, a new space, where knowledge can be shared and generations come together.

Music has also been used in more-than-representational health geographies in a number of ways. Engaging musical cultures, Skinner and Masuda (2014), for example, map health inequality in an urban area through participatory aboriginal hip hop. In this research young people 'mapped their city' in a series of workshops and events. Specifically, this involved them articulating their uses, experiences, feelings, attachments and contests across city space employing hip hop (its full spectrum of expressions: rap, art, dance) to depict their multisensory lived experiences in a multisensory way (involving shapes, colours, sounds, rhythms, movement and images). In another study, Andrews and Drass (2016) describe a collaboration between a geographer (Andrews) and an artist/producer of electronic dance music (Drass) to put affects and meanings into the world through two musical tracks: 'The Pump', a high-energy industrial techno track providing commentary and animation of bodybuilding and gym culture, and 'Senescence', an ambient track utilizing samples of body sounds as instrumentation to animate the body breaking down with age. The authors describe the 'Senescence' track:

> The track is slow, ambient and electronic opening with a first phase (0 to 1:15) composed of the breathing in and out of a body, a heart-beat beat, a haunting piano playing sparingly in the background, and regular bodily gurgles. After a lead-in (1:50–2:10), the second phase (2:10–2:50) increases in tempo notably. Metallic ticking sounds replace the heart-beat beat, and stressful in-time grunts and other percussive samples emerge. This second phase is faster and more frantic, but is soon over. After another short lead-in (2:50–3:20), a third phase (3:20–5:09) then slows everything down once again – but even slower than the fist – acting like a prolonged fade-out. The heart-beat beat returns but accompanied with complex fracturing of rhythm and broken piano melody. The song, reflecting its title, is a mid-life audit made purely from samples of Eric's own body with an artificial sub-base. The three musical phases reflect three phases of a body's life: (i) normal, youthful, stress-free with optimal cell replication; (ii) mid-life, hard work with consequences in terms of stress on the body and mind; (iii) later life, less stress with the replication of cells broken as the body breaks down.
>
> (2016:221)

Reflecting on the song's circulation (what one might think of traditionally as knowledge translation), the authors describe how an MP3 of the track was posted on Drass' personal and professional website and on YouTube and how, in the four years since, it has been downloaded hundreds of times. The authors also describe how Drass' website includes a personal/academic narrative about the track and wellbeing more broadly, complete with illustrations. They note that mirroring the track, this information is provocative rather than fully explanatory, leading readers to the door of diverse literatures and debates including on cell biology, music and time, emotion and soul, and rhythm in humans and social life.

As part of this project, Drass and Andrews also published a reflection on praxis (Andrews and Drass, 2016). Here they described the challenges that emerged under two themes. First, there was the challenge of doing NRT itself; of overcoming the academic urge to 'dig down' and theorize phenomena, and instead engage with life more lightly and directly. Second, there was the challenge of working with a very different type of professional and their medium (a researcher with an artist and their sounds, and an artist with an academic and their academic words). Indeed, they suggested that fitting music into scholarship and scholarship into music was not problem free, and that understanding one another and one another's motivations and approaches was an issue. Such collaboration, they suggested, can be tricky and conflict with personal and professional conventions, but the result was, for them at least, rewarding, bold and impactful.

Elsewhere, McCormack's research on dance movement therapy (DMT) provides insights into the multifaceted objectives, challenges and outcomes of engaging with art. The empirics of McCormack's (2003) project were, in part, designed to illustrate a particular theoretical point about progress in health geography (specifically that 'therapeutic landscapes' are not always revealed through linguistic signification). In making this point, however, McCormack through his participant observation, revealed so much more about the nature of therapeutic movements and the nature of researching them, which required flexibility and openness to change:

> [W]hat one folded in to an encounter with DMT was as important as what one found out. But this in turn meant that I also began to be less concerned with the effort to do justice to the experience of other participants in DMT through providing representations of, or 'giving voice' to their experiences. Indeed, in the case of an encounter with the often wordless intensities of DMT, it seemed that an effort to do justice to the thinking or inner experience of the participants was precisely what prevented one from becoming faithful to the relations and movements that played out through the enactment of the practice. Furthermore, as my aim shifted away from trying to 'give voice' to or 'represent' the experiences of the participants, I also felt less and less inclined to declare my 'position' as a researcher to anyone and everyone I encountered at DMT sessions.
>
> (2003:493)

> [O]ver time, my attention was drawn to the way in which the affective capacity of what happened during DMT sessions was always more than personal or indeed interpersonal experience. It was drawn to the pre-reflective and non-human dimensions of the relations and movement catalysed in the space of DMT. With this shift in attention, the 'ethics', or more precisely the ethos, of my encounter with DMT also shifted. It shifted away from figuring out how to extract information from the practice by understanding how its enactment represented hidden interests, internal or external. And it shifted towards an effort to cultivate a way of attending to and attending through the movements and relations that gave consistency to this practice.
>
> (2003:494)

In another study Boyd (2017) considers therapeutic art making. Her extensive ethnography involved five central practices: (i) recreating and re-practising some of the painted art she had produced in past years for her own benefit; (ii) travelling and working with therapeutic artists; (iii) involvement in dance movement therapy; (iv) taking part in an urban play; (v) involvement in fibre art (sewing). Boyd reflects on the research process which, for her, involved numerous unexpected possibilities and changes of directions, many of which, although transformative, were accidental and required that she allow herself to get lost in the art, its movement and potential. This, she reflects, included many breaches of typical research conventions:

> As an adjunct to this fieldwork, I collected numerous 'sense recordings' – video, photographs, audio – which may have served as data had a qualitative approach been employed. Instead, some of these recordings formed the basis of a body of creative work, which was formally exhibited. The majority, however, served only as 'tools for thought', held in memory until they might be re-imagined at the time of writing. This approach to fieldwork differs even further from ethnography in terms of operative notions of 'self as researcher', the failure to expressly position myself in relation to 'participants', and an eschewal of reflexive note taking…. The most radical breach, relates to data analysis or lack thereof… After Deleuze, writing can be considered as a performance that is only one small part of a wider extra-textual practice. In this vein, my 'findings' have four alternate forms, each asserting the primacy of practice – an audio/visual form, an exhibited form, a poetic form, and a ghostly form… The ghostly form resides in the past, never to be presented, or is hidden away in various places on the Internet.
>
> (2017:44–45)

New styles and forms of writing and presenting

Writing and presenting more-than-representational health geographies has taken many forms, often adopting the irrealis or 'play-by-play' style, noted previously, to bring to life the events researched. In his edited collection on methods in NRT, Vannini (2015c) introduces Lingis' (2015) work that narrates 'momentous journeys' (year-long bike rides and walking long distances from city to city) to consecrate personal losses. Vannini reflects that Lingis' writing, like the loss he conveys, leaves voids and after-images that are meant to 'ring hollow' with readers, conveying a sense of nothingness, and showing nothing other than the will to make the journey (for readers, the barren lines of his text taking them on a similar emotionless yet sensory journey). An extract from Lingis' chapter reads:

> Michael had no cause, no goal or purpose. It was not something he was doing for Kelly; Kelly was lost forever in the black void. To pump the bike for 10,000 kilometers makes one completely physical. Consciousness exists now in the tensions and the relaxing of the muscles, in the feeling of strength and in the fatigue. Consciousness exists on the surfaces of the body all sensitive to the sun, the wind, the cold, the rain. A consciousness that excludes thinking, remembering, envisioning works and ambitions. The road rising and descending, kilometer after kilometer. Sometimes the landscape opens upon enchanting vistas, glistening with dew and birdsong. Sometimes physical fatigue blurs the eyes, the landscape dissolves

into green dust. There is no planning the day ahead; who knows what the weather will be, what the road will be. The end of the day one sinks heavily into dark sleep.... The open road drew movement into him. During the 10,000 kilometers, nature was tunneling into him in his strong breathing, strong pushing, strong feeling, strong forcing, strong dancing, strong singing-out. He felt nature guiding his body and felt an intensity of trust that he had never known before. The sun and the breeze fueled his body. He was a body in nature, like a hare in the prairie, a bird in the sky. Unbounded nature, horizons opening endlessly onto more nature.

(2015:216–217)

Taking a slightly different approach, Anderson (2014) considers writing as data. His examination of popular surf writing (typically focused on surfing as a joyful, spiritual wellness experience) discusses its ability to fill what he describes as the 'intersubjective space' between writer and reader – helping overcome problems with articulation and representation. Indeed, Anderson suggests that, just like academic writers, surf writers need to 'get at' the world that lies beyond words. In attempting to do this, they seek to connect to readers' own previous affective experiences to create a resonance with them, the surf writing evoking both physical and emotional engagement. Thus a common understanding – and feeling – is reached in the 'space between' writer and reader (evoking sensation and emotion in this case being very different than 'explaining' emotion; the latter oftentimes resulting in frustration both for the writer and reader due to the continued absence of that sensation and emotion). Evoking the space between writer and reader, Anderson argues, might be one way to overcome the paradox of non-representation in NRT (as in this space-between there is no representation going on). The surf writing Anderson reviews (and shows) might, for example, take a spectator's perspective:

From the granite headland whose rocks were daubed with warnings about the dangerous current, the beach stretched east for miles. We watched the surfers plunge into a churning rip alongside the rocks and from there they shot out towards the break. Waves ground around the headland, line upon line of them, smooth and turquoise, reeling across the bay to spend themselves in a final mauling rush against the bar at the river mouth. The air seethed with noise and salt; I was giddy with it.

(2014:31)

or a participant's:

Although the swell had dropped a bit, it was still booming. The set waves were well overhead, burnished smooth by a light offshore wind. It occurred to me how deceiving the wave at Salsa is from the distant, shore-side view. [Did] you truly have to be there, in the pit, to understand?

(2014:31)

Notably, in many studies in more-than-representational health geography academic writing – no matter how expressive – is combined with photos, images, diagrams and other things to show movement, and to provide different access points to the activity and event. McCormack (2013), for example, uses this to good effect (affect) in his studies of dance movement. Such images can also be more fully part of the research process itself. For example, as part of Ravn and Duff's (2015) aforementioned study of drug use in private house parties, the authors introduced a 'map task' to complement other fieldwork and get to these hidden spatialities. Specifically, respondents drew a map of a party they had recently attended, and the researcher asked them questions during this task about what went on and when. The authors noted that they found this approach to be useful for showing movement through space and time, and less-than-fully consciously practised spatial routines.

Unpacking and changing assemblages for health: towards a new ethics

There is certainly a need to think of some ways forward for NRT in health geography, within and beyond the varied methodological engagements of more-than-representational health geographies discussed thus far. With this in mind it is worth briefly returning to the fundamental ideas on health presented by Duff (2014a). As Duff suggests, individuals are never permanently sick or well, but traverse a line between sick and well; a line of becoming that is dependent on the assemblages they are exposed to and part of and their affects, events and relations (this contrasting with a traditional view in the health sciences that health is a result of what is structurally done to individuals or what individuals do to themselves). As Duff argues then, and as suggested in Chapter 2, part of a new ethology of health under NRT will involve determining the components of assemblages from the trails they leave, and their various roles and contributions in creating or degrading health (see also Fox, 2011). In other words, methodologies and projects are needed that can isolate and describe causal pathways that underpin health's emergence, key questions (as Duff suggests) including: (i) What assemblages provide the richest and best experiences (affectionate, therapeutic, supportive etc)? (ii) How are assemblages and their encounters organized to create situations whereby bodies are in full possession of their power of acting? (iii) Connecting to existing post-structuralist concerns in research; which forms of pre-existing political, economic, institutional and social processes help or hinder assemblages for health – which should be used, which should be resisted, which should be ignored? As Greenhough (2011b) suggests, these questions and their related projects help pull away from binary positions and stereotypes articulated in much health geography and even critical health geography (e.g. paternalistic, objective biomedicine imposing itself on vulnerable, passive patients who might resist) by realizing the numerous dynamics and inputs at play in continuing processes that co-create overall health situations. This is a perspective based in 'real' happenings rather than on constructs.

Notably, Duff suggests that such an agenda is underpinned by a new ethical position; a focus on the 'ethics of assemblage'. Indeed, whereas traditional ethics of health care is primarily concerned with encouraging the best biomedical attempts to deliver individuals and populations to pre-morbid states, what matters more ethically to NRT is what individuals can draw upon to maintain their recovery or a certain level of health (Duff, 2014a). Thus, this new ethical position, in understanding the potential and capacities of bodies and objects, asks what else they 'can do' in encounters with each other to increase health. Moreover it asks what can also 'be done' to enhance their capacities and potential (thus, ethically, health is understood as a malleable relational mode of existence rather than a psychological or physiological pre-condition). As Duff suggests, deciding what encounters might be more unhealthy or healthy requires ethical judgement, not based on transcendental rules or power, but based on a belief in the plane of immanence and life as a self-organizing process. As McCormack (2003) suggests, this new ethics moves beyond traditional judgements of what should/should not happen or should be/should not be happening regarding health according to codified rules, and involves attending directly to the forces circulated by capitalism and the state. It involves a realization that 'common sense' practice emerges from focusing on affective visceral, less-than-fully conscious processes, in thinking what might be good or bad or about them, and in decisions whether to intervene affectively to change the affective powers of spaces. As McCormack (2003) suggests, all this involves a new ethical sensibility on the part of researchers involving knowing this and being attentive to it in action:

> The importance of this is not merely to extend the ethical entanglements of care and responsibility to non-human agents in ways that are haunted by the question of how close such agents arc to the human. Rather, what is involved here is the effort to energize ethics by admitting that the corporeal finitude of the human is emergent from a connective multiplicity of non-human and in-human forces and processes that exceed this corporeality in extensive, intensive, temporal and ontogenetic ways.... The important point here is that if much of the world is emergent from the processual enactment of non-representational processes then it is surely imperative not to limit the field of the ethical to judgements made upon the basis of already articulated codes.
>
> (2003:489)

This new ethics also suggests that a potential educational enterprise could exist with and through research. Indeed, as Duff (2014a) posits, if we believe – like Deleuze (1992) – that that the body is 'naturally' inclined to seek what is good for it, it follows that researchers should consciously strive to do the same: to unite the body with its constructive milieus. Through this education, greater self-awareness and self-reflection can occur as to what assemblages and how assemblages create health or ill-health. Through this self-awareness and self-reflection change might occur when, as Duff argues, (i) institutions see how they contribute to health creating or detracting processes and alter their behaviours, and (ii) individuals and groups seek to change the character of their encounters. In all this we need to avoid implementing yet more rules, codes and regulations regarding what to do and what not to do with respect to health. Perhaps instead, as Duff notes, we might openly 'play with forces' – thinking about which ones can be used, which ones can be moulded, and which ones can be avoided?

Concluding thoughts on other ways forward...

One immediate way forward for NRT in health geography is to, by example, directly address the concerns that doubters have over the approach. Indeed, as outlined in Chapter 1, some scholars have taken issue with what they see as the universalist nature of NRT particularly in terms of its relationality. It, they argue, fails to differentiate bodies and recognize persons through social and demographic categories that impact upon people and consti- tute the ways in which they understand themselves and their lives (Bondi, 2005). Moreover, critics argue that NRT fails to recognize political power (Jacobs and Nash, 2003; Pain, 2006), is technocratic and abstract (Thien, 2005; Bondi, 2005), and distances deep feelings and emotions (Thien, 2005). I consider there to be two reasons, however, why these problems and shortfalls need not arise: because strong arguments are emerging from scholars as to how NRT can expose why certain bodies are marginalized and the non-representational processes involved (Boyer, 2012; Colls, 2012b; Hall and Wilton, 2017), and because writings on NRT and health have never claimed that existing research diversity, including emotion-sensitive approaches, should be shelved (they recognizing that there is plenty of room for both the 'old' and 'new' in health geography). Like Colls (2012b), I see potential in developing a 'nomadic consciousness' in research that is able to roam between the representational and non-representational. This approach – applicable both within single studies and between studies within wider fields of scholarship – simply reflects the reality that in life much that is active, often less-than-fully conscious and non-representational eventu- ally flows into the conscious representable realm where power, meaning, identity and such things are more clearly verbalized.

Evidently, more-than-representational health geography is expanding, but it is currently a little disparate as a body of work, lacking overall cohesion and direction (beyond an exciting early aim to expose health's immediacies). With this is mind, a number of broader questions and priorities exist for the sub-discipline. First, empirical gaps remain in the literature that need to be filled. Although empirical coverage is ever broadening, there seems to be an abundance of work, for example, in areas such as wellbeing, movement activities, CAM and urban environments, but a relative paucity of work on conventional health and social care, public health initiatives, environmental health, living with illness, and settings such as homes, hospitals and rural environments. The agenda here should not however be simply to reach blanket coverage of health and health care through NRT. Hence, second, we need to consider what are the most important and pressing contemporary health issues and trends that NRT could help expose and ultimately address. Indeed, we need to think about what the particular contribution of NRT-informed health geography might be to what particular service, policy and practice matters. Third, and related, we need to think about the academic contribution of NRT to health geography and health research more broadly. To which concepts, fields and debates might it contribute the most, adding what knowledge specifically? Fourth, we need to think about the development of NRT within health geography itself. Beyond current more-than-representational health geography – that could be considered by purists to be a little 'NRT lite' – we need to develop more of an explicit NRT approach involving its own ideas, perhaps eventually even making contributions back to NRT in human geography as a whole. Fifth, and finally, we need to think more about which particular NRT-derived methods fit which particular health contexts and research questions and how we might introduce methodological innovations towards the ethical goal of working with, and changing, assemblages for health (as discussed above).

Health geography has always lagged a little way behind the parent discipline in terms of theoretical innovation and progress. However, it has never ignored latest developments and has always caught onto them eventually, on each occasion adding to the richness of its scholarship and the range of its windows onto the world. As the data mined and conveyed in this book shows, in the case of NRT, the sub-discipline might not be too far behind the curve, with clear early precedent existing across a number of popular subject fields. Hence the five questions noted above are important to furthering NRT in health geography and the ontological turn it offers the sub-discipline.

References

Adams, A., Theodore, D., Goldenberg, E., McLaren, C., & McKeever, P. (2010). Kids in the atrium: comparing architectural intentions and children's experiences in a pediatric hospital lobby. *Social Science & Medicine*, *70*(5), 658–667.

Adler, P., & Adler, P. A. (1978). The role of momentum in sport. *Journal of Contemporary Ethnography*, *7*(2), 153–175.

Affonso, D. D., Andrews, G. J., & Jeffs, L. (2004). The urban geography of SARS: Paradoxes and dilemmas in Toronto's health care. *Journal of Advanced Nursing*, *45*(6), 568–578.

Ali, S. H., & Keil, R. (2007). Contagious cities. *Geography Compass*, *1*(5), 1207–1226.

Amin, A. (2004). Regions unbound: Towards a new politics of place. *Geografiska Annaler: Series B, Human Geography*, *86*(1), 33–44.

Anderson, B. (2002). A principle of hope: Recorded music, listening practices and the immanence of utopia. *Geografiska Annaler: Series B, Human Geography*, *84*(3–4): 211–227.

Anderson, B. (2006). Becoming and being hopeful: Towards a theory of affect. *Environment and Planning D*, *24*(5), 733–752.

Anderson, B. (2007). Hope for nanotechnology: Anticipatory knowledge and the governance of affect. *Area*, *39*(2), 156–165.

Anderson, B. (2009). Affective atmospheres. *Emotion, Space and Society*, *2*(2), 77–81.

Anderson, B. (2010). Preemption, precaution, preparedness: Anticipatory action and future geographies. *Progress in Human Geography*, *34*(6), 777–798.

Anderson, B. (2012). Affect and biopower: Towards a politics of life. *Transactions of the Institute of British Geographers*, *37*(1), 28–43.

Anderson, B., Harrison, P. (2010). *Taking place: Non-representational theories and geography*. Ashgate.

Anderson, J. (2014). Exploring the space between words and meaning: Understanding the relational sensibility of surf spaces. *Emotion, Space and Society*, *10*(1), 27–34.

Andrews, G. J. (2003). Placing the consumption of private complementary medicine: Everyday geographies of older peoples' use. *Health & Place*, *9*(4), 337–349.

Andrews, G. J. (2004). (Re)thinking the dynamic between healthcare and place: Therapeutic geographies in treatment and care practices. *Area*, *36*(3), 307–318.

Andrews, G. J. (2006). Geographies of health in nursing. *Health & Place*, *12*(1), 110–118.

Andrews, G. J. (2007). Spaces of dizziness and dread: Navigating acrophobia. *Geografiska Annaler: Series B, Human Geography*, *89*(4), 307–317.

Andrews, G. J. (2011). 'I had to go to the hospital and it was freaking me out': Needle phobic encounter space. *Health & Place*, *17*(4), 875–884.

Andrews, G. J. (2014a). Co-creating health's lively, moving frontiers: Brief observations on the facets and possibilities of non-representational theory. *Health & Place*, *30*, 165–170.

Andrews, G. J. (2014b). 'Gonna live forever': Noel Gallagher's spaces of wellbeing. In Andrews, G. J., Kingsbury, P., Kearns, R. A. (Eds.), *Soundscapes of and wellbeing in popular music*. Ashgate.

Andrews, G. J. (2014c). A force from the beginning: Wellbeing in the affective intensities of pop music. *Aporia*, *6*(4), 6–8.

Andrews, G. J. (2015). The lively challenges and opportunities of non-representational theory: A reply to Hanlon and Kearns. *Social Science & Medicine*, *128*, 338–341.

Andrews, G. J. (2016). Geographical thinking in nursing inquiry, part one: Locations, contents, meanings. *Nursing Philosophy*, *17*(4), 262–281.

Andrews, G. J. (2017a). From post-game to play-by-play: Animating sports movement-space. *Progress in Human Geography*, (in press) 0309132516660207.

Andrews, G. J. (2017b). 'Running hot': Placing health in the life and course of the vital city. *Social Science & Medicine*, *175*, 209–214.

Andrews, G. J. (2018). New brain geographies: Living with Chiari malformation. *Social Science and Medicine* (in press).

Andrews, J. P., & Andrews, G. J. (2003). Life in a secure unit: The rehabilitation of young people through the use of sport. *Social Science & Medicine 56*, (3), 531–550.

Andrews, G. J., & Chen, S. (2006). The production of tyrannical space. *Children's Geographies*, *4*(2), 239–250.

Andrews, G. J., & Drass, E. (2016). From 'The pump' to 'Senescence'. In Fenton, N. E., & Baxter, J. (Eds.), *Practicing qualitative methods in health geographies*. Routledge.

Andrews, G. J., & Evans, J. (2008). Understanding the reproduction of health care: Towards geographies in health care work. *Progress in Human Geography*, *32*(6), 759–780.

Andrews, G. J., & Grenier, A. M. (2015). Ageing movement as space-time: Introducing non-representational theory to the geography of ageing. *Progress in Geography* (Chinese), *34*(12), 1512–1534.

Andrews, G. J., & Holmes, D. (2007). Gay bathhouses: Transgressions of health in therapeutic places. In Williams, A. (Ed.), *Therapeutic landscapes*. Ashgate.

Andrews, G. J., & Kearns, R. A. (2005). Everyday health histories and the making of place: The case of an English coastal town. *Social Science & Medicine*, *60*(12), 2697–2713.

Andrews, G. J., & Phillips, D. R. (2002). Changing local geographies of private residential care for older people 1983–1999: Lessons for social policy in England and Wales. *Social Science & Medicine*, *55*(1), 63–78.

Andrews, G. J., & Shaw, D. (2008). Clinical geography: Nursing practice and the (re) making of institutional space. *Journal of Nursing Management*, *16*(4), 463–473.

Andrews, G. J., & Shaw, D. (2010). 'So we started talking about a beach in Barbados': Visualization practices and needle phobia. *Social Science & Medicine*, *71*, 1804–1810.

Andrews, G. J., & Shaw, D. (2012). Place visualization: Conventional or unconventional practice? *Complementary Therapies in Clinical Practice*, *18*(1), 43–48.

Andrews, G. J., Sudwell, M. I., & Sparkes, A. C. (2005). Towards a geography of fitness: An ethnographic case study of the gym in British bodybuilding culture. *Social Science & Medicine, 60*(4), 877–891.

Andrews, G. J., Kearns, R. A., Kingsbury, P., & Carr, E. R. (2011). Cool aid? Health, wellbeing and place in the work of Bono and U2. *Health & Place, 17*(1), 185–194.

Andrews, G. J., Hall E., Evans, B., & Colls, R. (2012). Moving beyond walkability: On the potential of health geography. *Social Science & Medicine, 75*(11), 1925–1932.

Andrews, G. J., Evans J., & McAlister, S. (2013). 'Creating the right therapy vibe': Relational performances in holistic medicine. *Social Science & Medicine, 83*, 99–109.

Andrews G. J., Chen S., & Myers S. (2014). The 'taking place' of health and wellbeing: Towards non-representational theory. *Social Science & Medicine, 108*, 210–222.

Antoninetti, M., & Garrett, M. (2012). Body capital and the geography of aging. *Area, 44*(3), 364–370.

Aslam, A., & Corrado, L. (2011). The geography of well-being. *Journal of Economic Geography, 12*(3), 627–649.

Atkinson, S. (2013). Beyond components of wellbeing: The effects of relational and situated assemblage. *Topoi, 32*(2), 137.

Atkinson, S. J., & Joyce, K. E. (2011). The place and practices of wellbeing in local governance. *Environment and Planning C: Government and Policy, 29*, 133–148.

Atkinson, S., & Rubidge, T. (2013). Managing the spatialities of arts-based practices with school children: An inter-disciplinary exploration of engagement, movement and well-being. *Arts & Health, 5*(1), 39–50.

Atkinson, S. J., & Scott, K. E. (2015). Stable and destabilised states of subjective wellbeing: Dance and movement as catalysts of transition. *Social & Cultural Geography, 16*, 75–94.

Atkinson, S., Fuller, S., & Painter, J. (Eds.). (2012). *Wellbeing and place*. Ashgate.

Baer, L. D. (2003). A proposed framework for analyzing the potential replacement of international medical graduates. *Health & Place, 9*(4), 291–304.

Baer, L. D., & Gesler, W. M. (2004). Reconsidering the concept of therapeutic landscapes in J. D. Salinger's *The Catcher in the Rye. Area, 36*(4), 404–413.

Bailey, A J. (2009). Population geography: Lifecourse matters. *Progress in Human Geography, 33*(3), 407–415.

Bale, J. (2004). *Running cultures: Racing in time and space*. Routledge.

Barnes, L., & Rudge, T. (2005). Virtual reality or real virtuality: The space of flows and nursing practice. *Nursing Inquiry, 12*(4), 306–315.

Barnett, J. R. (1988). Foreign medical graduates in New Zealand 1973–1979: A test of the 'exacerbation hypothesis'. *Social Science & Medicine, 26*(10), 1049–1060.

Barnett, J. R. (1991). Geographical implications of restricting foreign medical immigration: A New Zealand case study, 1976–1987. *Social Science & Medicine, 33*(4), 459–470.

Barnett, J. R., & Barnett, P. (2003). 'If you want to sit on your butts you'll get nothing!' Community activism in response to threats of rural hospital closure in southern New Zealand. *Health & Place, 9*(2), 59–71.

Barnfield, A. (2016). Public health, physical exercise and non-representational theory: A mixed method study of recreational running in Sofia, Bulgaria. *Critical Public Health, 26*(3), 281–293.

Barratt, P. (2011). Vertical worlds: Technology, hybridity and the climbing body. *Social & Cultural Geography, 12*, 397–412.

Barratt, P. (2012). 'My magic cam': a more-than-representational account of the climbing assemblage. *Area 44*, 1, 46–53.

Beck, H. (2012). Understanding the impact of urban green space on health and well-being. In Atkinson, S., Fuller, S., & Painter, J. (Eds.). (2012). *Wellbeing and place*. Ashgate.

Bell, S. L., Phoenix, C., Lovell, R., & Wheeler, B. W. (2015). Using GPS and geo-narratives: A methodological approach for understanding and situating everyday green space encounters. *Area, 47*(1), 88–96.

Bender, A., Andrews, G. J., Peter, E. (2010). Displacement and tuberculosis: Recognition in nursing care. *Health & Place, 16*(6), 1069–1076.

Bennett, J. (2009). *Vibrant matter: A political ecology of things*. Duke University Press.

Bentham, G. (1988). Migration and morbidity: Implications for geographical studies of disease. *Social Science & Medicine, 26*(1), 49–54.

Bergmann, S. (2008). *The beauty of speed or the discovery of slowness: Why do we need to rethink mobility?* In Bergmann, S., & Sager, T. (Eds.), *The ethics of mobilities: Rethinking place, exclusion, freedom and environment*. Ashgate.

Bergson, H. (1889). *Time and free will*. (F. L. Pogson, Trans.). London: Sonnenschein.

Bergson, H. (1911a). *Matter and memory*. (N.M. Paul, W.S. Palmer, Trans.). Macmillan. (Orig. pub. 1896).

Bergson, H. (1911b). *Creative evolution*. (A. Mitchell, Trans.). Macmillan. (Orig. pub. 1907).

Bissell, D. (2008). Comfortable bodies: Sedentary affects. *Environment and Planning A, 40*(7), 1697–1712.

Bissell, D. (2009). Obdurate pains, transient intensities: Affect and the chronically pained body. *Environment and Planning A, 41*(4), 911–928.

Bissell, D. (2010). *Placing affective relations: Uncertain geographies of pain*. In Anderson, B., & Harrison, P. (Eds.), *Taking place: Non-representational theories and geography*. Ashgate, 79–98.

Bissell, D. (2011). Thinking habits for uncertain subjects: movement, stillness, susceptibility. *Environment and Planning A, 43*(11), 2649–2665.

Bissell, D. (2012). Agitating the powers of habit: Towards a volatile politics of thought. *Theory & Event, 15*(1). http://muse.jhu.edu/article/469326.

Bissell, D. (2013). Habit displaced: The disruption of skilful performance. *Geographical Research, 51*(2), 120–129.

Bissell, D. (2015). Virtual infrastructures of habit: The changing intensities of habit through gracefulness, restlessness and clumsiness. *Cultural Geographies, 22*(1), 127–146.

Bissell, D., & Fuller, G. (Eds.). (2011). *Stillness in a mobile world*. Routledge.

Bøhling, F. (2014). Crowded contexts: On the affective dynamics of alcohol and other drug use in nightlife spaces. *Contemporary Drug Problems, 41*(3), 361–392.

Bondi, L. (2005). Making connections and thinking through emotions: Between geography and psychotherapy. *Transactions of the Institute of British Geographers, 30*, 433–488.

Boyd, C. P. (2017). *Non-representational geographies of therapeutic art making: Thinking through practice*. Palgrave Macmillan.

Boyer, K. (2011). 'The way to break the taboo is to do the taboo thing': Breastfeeding in public and citizen-activism in the UK. *Health & Place, 17*(2), 430–437.

Boyer, K. (2012). Affect, corporeality and the limits of belonging: Breastfeeding in public in the contemporary UK. *Health & Place, 18*(3), 552–560.

Boyle, P. (2004). Population geography: Migration and inequalities in mortality and morbidity. *Progress in Human Geography*, *28*(6), 767–776.

Braun, B. (2007), Biopolitics and the molecularization of life. *Cultural Geographies*, *14*(1), 6–28.

Briggs, D. J., Collins, S., Elliott, P., Fischer, P., Kingham, S., Lebret, E., ... Van Der Veen, A. (1997). Mapping urban air pollution using GIS: A regression-based approach. *International Journal of Geographical Information Science*, *11*(7), 699–718.

Brimblecombe, N., Dorling, D., & Shaw, M. (2000). Migration and geographical inequalities in health in Britain. *Social Science & Medicine*, *50*(6), 861–878.

Brodie, D. A., Andrews, G. J., Andrews, J. P., Thomas, B. G., Wong J., & Rixon, L. (2005). Working in London hospitals: Perceptions of place in nursing students' employment considerations. *Social Science & Medicine 61*(9), 1867–1881.

Brown, T. (2003). Towards an understanding of local protest: Hospital closure and community resistance. *Social & Cultural Geography*, *4*(4), 489–506.

Brown, K. M. (2017). The haptic pleasures of ground-feel: The role of textured terrain in motivating regular exercise. *Health & Place*, *46*, 307–314.

Brown, T., & Bell, M. (2007). Off the couch and on the move: Global public health and the medicalization of nature. *Social Science & Medicine*, *64*, 1343–1354.

Brown, T., & Bell, M. (2008). Imperial or postcolonial governance? Dissecting the genealogy of a global public health strategy. *Social Science & Medicine*, *67*, 1571–1579.

Brown, T., & Burges Watson, D. L. (2010). Governing un/healthy populations. In Brown, T., McLafferty, S., & Moon, G. (Eds.), *A companion to health and medical geography*. Wiley-Blackwell, 477–494.

Brown, T., & Duncan, C. (2002). Placing geographies of public health. *Area*, *34*(4), 361–369.

Brown, S. D., & Middleton, D. (2005). The baby as virtual object: Agency and difference in a neonatal intensive care unit. *Environment and Planning D*, *23*(5), 695.

Brown, B. B., Yamada, I., Smith, K. R., Zick, C. D., Kowaleski-Jones, L., & Fan, J. X. (2009). Mixed land use and walkability: Variations in land use measures and relationships with BMI, overweight and obesity. *Health & Place*, *15*(4), 1130–1141.

Brown, T., Craddock, S., & Ingram, A. (2012). Critical interventions in global health: Governmentality, risk, and assemblage. *Annals of the Association of American Geographers*, *102*(5), 1182–1189.

Buchanan, I. (2000). *Deleuzism: A metacommentary*. Duke University Press.

Burges Watson, D., Murtagh, M. J., Lally, J. E., Thomson, R. G., & McPhail, S. (2007). Flexible therapeutic landscapes of labour and the place of pain relief. *Health & Place*, *13*(4), 865–876.

Cadman L. (2009). Nonrepresentational theory/nonrepresentational geographies. In Kitchin, R., & Thrift, N. (Eds.), *International encyclopedia of human geography*. Elsevier.

Carolan, M. S. (2008). More-than-representational knowledge/s of the countryside: How we think as bodies. *Sociologia Ruralis*, *48*(4), 408–422.

Carolan, M., Andrews G. J., & Hodnett, E. (2006). Writing place: A comparison of nursing research and health geography. *Nursing Inquiry*, *13*(3), 203–219.

Carr-Hill, R. A., Sheldon, T. A., Smith, P., Martin, S., Peacock, S., & Hardman, G. (1994). Allocating resources to health authorities: development of method for small area analysis of use of inpatient services. *BMJ: British Medical Journal*, *309*(6961), 1046.

Cheek, J. (2004). Older people and acute care: A matter of place. *Illness, Crisis and Loss, 12*(1), 52–62.

Cho, S. H., Lee, J. Y., Mark, B. A., & Jones, C. B. (2014). Geographic mobility of Korean new graduate nurses from their first to subsequent jobs and metropolitan–nonmetropolitan differences in their job satisfaction. *Nursing Outlook, 62*(1), 22–28.

Chordia, T., & Shivakumar, L. (2002). Momentum, business cycle, and time-varying expected returns. *The Journal of Finance, 57*(2), 985–1019.

Church, R. L., & Meadows, M. E. (1979). Location modeling utilizing maximum service distance criteria. *Geographical Analysis, 11*(4), 358–373.

Clarke, D. (1973). Archaeology: The loss of innocence. *Antiquity, 47*, 6–18.

Clarke, D. B., Doel, M. A., & Segrott, J. (2004). No alternative? The regulation and professionalization of complementary and alternative medicine in the United Kingdom. *Health & Place, 10*(4), 329–338.

Cloutier-Fisher, D., & Skinner, M. W. (2006). Levelling the playing field? Exploring the implications of managed competition for voluntary sector providers of long-term care in small town Ontario. *Health & Place, 12*(1), 97–109.

Colby, C. C. (1933). Centrifugal and centripetal forces in urban geography. *Annals of the Association of American Geographers, 23*(1), 1–20.

Coleman, T., & Kearns, R. (2015). The role of bluespaces in experiencing place, aging and wellbeing: Insights from Waiheke Island, New Zealand. *Health & Place, 35*, 206–217.

Colls, R. (2007). Materialising bodily matter: Intra-action and the embodiment of 'fat'. *Geoforum, 38*(2), 353–365.

Colls, R. (2012a). Big girls having fun: Reflections on a 'fat accepting space'. *Somatechnics, 2*(1), 18–37.

Colls, R. (2012b). Feminism, bodily difference and non-representational geographies. *Transactions of the Institute of British Geographers, 37*(3), 430 445.

Connell, J., & Walton-Roberts, M. (2016). What about the workers? The missing geographies of health care. *Progress in Human Geography, 40*(2), 158–176.

Conradson, D. (2003). Spaces of care in the city: The place of a community drop-in centre. *Social & Cultural Geography, 4*(4), 507–525.

Conradson, D. (2005). Freedom, space and perspective: Moving encounters with other ecologies. In Davidson, J., Bondi, L., & Smith, M. (Eds.), *Emotional geographies*. Ashgate, 103–116.

Conradson, D. (2005b). Landscape, care and the relational self: Therapeutic encounters in rural England. *Health & Place, 11*(4), 337–348.

Conradson, D. (2007). The experiential economy of stillness: Places of retreat in contemporary Britain. In Williams, A. (Ed.), *Therapeutic landscapes*. Ashgate, 33–47.

Conradson, D. (2011). The orchestration of feeling: stillness, spirituality and places of retreat. In Bissell, D., & Fuller, G. (Eds.), *Stillness in a mobile world*. Routledge, S, 71–86.

Conradson, D., & Latham, A. (2007). The affective possibilities of London: Antipodean transnationals and the overseas experience. *Mobilities, 2*(2), 231–254.

Conradson, D., & Moon, G. (2009). On the street: Primary health care for difficult to reach populations. Crooks, V., & Andrews, G. J. (Eds.), *Primary health care: People, practice, place*. Ashgate.

Cook, M., & Edensor, T. (2017). Cycling through dark space: Apprehending landscape otherwise. *Mobilities, 12*(1), 1–19.

Cook, S., Shaw, J., & Simpson, P. (2016). Jography: Exploring meanings, experiences and spatialities of recreational road-running. *Mobilities*, 11(5), 744–769.

Cooper, M. J., Gutierrez, R. C., & Hameed, A. (2004). Market states and momentum. *The Journal of Finance*, 59(3), 1345–1365.

Corden, A. (1992). Geographical development of the long-term care market for elderly people. *Transactions of the Institute of British Geographers*, 1(1), 80–94.

Courtney, K. L. (2005). Visualizing nursing workforce distribution: Policy evaluation using geographic information systems. *International Journal of Medical Informatics* 74(11), 980–988.

Craddock, S. (1999). Embodying place: Pathologizing Chinese and Chinatown in nineteenth-century San Francisco. *Antipode*, 31(4), 351–371.

Craddock, S., & Brown, T. (2010). *Representing the (un)healthy body*. In Brown, T., McLafferty, S., & Moon, G. (Eds.), A companion to health and medical geography. John Wiley & Sons.

Cresswell, T. (2012). Nonrepresentational theory and me: Notes of an interested sceptic [Review essay]. *Environment and Planning D*, 30 (1), 96–105.

Cresswell, T., & Merriman, P. (Eds.). (2011). *Geographies of mobilities: Practices, spaces, subjects*. Ashgate.

Crighton, E. J., Upshur, R. E. G., & Mamdani, M. (2001). A population based time series analysis of asthma hospitalizations in Ontario, Canada: 1988 to 2000. *BMC Health Services Research*, 1(7), 1–6.

Crighton, E. J., Elliott, S. J., Kanaroglou, P., Moineddin, R., & Upshur, R. E. (2008). Spatio-temporal analysis of pneumonia and influenza hospitalizations in Ontario, Canada. *Geospatial Health*, 2(2), 191–202.

Crooks, V. A. (2007). Exploring the altered daily geographies and lifeworlds of women living with fibromyalgia syndrome: A mixed-method approach. *Social Science & Medicine*, 64(3), 577–588.

Crooks, V. A., & Chouinard, V. (2006). An embodied geography of disablement: Chronically ill women's struggles for enabling places in spaces of health care and daily life. *Health & Place*, 12(3), 345–352.

Crooks, V. A., & Evans, J. (2007). *The writing's on the wall: Decoding the interior space of the hospital waiting room*. In Williams, A. (Ed.), Therapeutic landscapes. Ashgate, 165–180.

Crooks, V. A., Dorn, M. L., & Wilton, R. D. (2008). Emerging scholarship in the geographies of disability. *Health & Place*, 14(4), 883–888.

Crooks, V. A., Turner, L., Snyder, J., Johnston, R., & Kingsbury, P. (2011). Promoting medical tourism to India: Messages, images, and the marketing of international patient travel. *Social Science & Medicine*, 72(5), 726–732.

Crossley, N. (1995). Merleau-Ponty, the elusive body and carnal sociology. *Body & Society*, 1(1), 43–63.

Crossley, N. (2004). The circuit trainer's habitus: Reflexive body techniques and the sociality of the workout. *Body & Society*, 10(1), 37–70.

Cummins, S., & Macintyre, S. (2006). Food environments and obesity: neighborhood or nation? *International Journal of Epidemiology*, 35, 100–104.

Cummins, S., Curtis, S., Diez-Roux, A., V., & Macintyre, S. (2007). Understanding and representing 'place'in health research: A relational approach. *Social Science & Medicine*, 65(9), 1825–1838.

Curtis, S., & Jones, I. (1998). Is there a place for geography in the analysis of health inequality? *Sociology of Health & Illness*, 20(5), 645–672.

Curtis, S., & Riva, M. (2010a). Health geographies I: Complexity theory and human health. *Progress in Human Geography, 34*(2) 215–223.

Curtis, S., & Riva, M. (2010b). Health geographies II: Complexity and health care systems and policy. *Progress in Human Geography, 34*(4), 513–520.

Curtis, S., Gesler, W., Fabian, K., Francis, S., & Priebe, S. (2007). Therapeutic landscapes in hospital design: A qualitative assessment by staff and service users of the design of a new mental health inpatient unit. *Environment and Planning C, 25*(4), 591.

Curtis, S., Gesler, W., Priebe, S., & Francis, S. (2009). New spaces of inpatient care for people with mental illness: A complex 'rebirth'of the clinic? *Health & Place, 15*(1), 340–348.

Cutchin, M. P. (1997). Physician retention in rural communities: The perspective of experiential place integration. *Health & Place, 3*(1), 25–41.

Cutchin, M. P. (1999). Qualitative explorations in health geography: Using pragmatism and related concepts as guides. *The Professional Geographer, 51*(2), 265–274.

Cutchin, M. P. (2003). The process of mediated aging-in-place: A theoretically and empirically based model. *Social Science & Medicine, 57*(6), 1077–1090.

Cutchin, M. P. (2004). A Deweyan case for the study of uncertainty in health geography. *Health & Place, 10*(3), 203–213.

Cutchin, M. P. (2008). John Dewey's metaphysical ground-map and its implications for geographical inquiry. *Geoforum, 39*(4), 1555–1569.

Darling, J. (2009). Thinking beyond place: The responsibilities of a relational spatial politics. *Geography Compass, 3*(5), 1938–1954.

Davidson, J. (2000). '... The world was getting smaller': Women, agoraphobia and bodily boundaries. *Area, 32*(1), 31–40.

Davidson, J. (2003). *Phobic geographies: The phenomenology and spatiality of identity.* Gower.

Davidson, J. (2005). Contesting stigma and contested emotions: Personal experience and public perception of specific phobias. *Social Science & Medicine, 61*(10), 2155–2164.

Dawney, L. (2013). The interruption: Investigating subjectivation and affect. *Environment and Planning D: Society and Space, 31*(4), 628–644.

Dean, J. (2016). Walking in their shoes: Utilizing go-along interviews to explore participant engagement with local space. In Fenton, N. E., & Baxter, J. (Eds.), *Practicing qualitative methods in health geographies.* Routledge.

Dear, M. J., & Wolch, J. R. (1987). *Landscapes of despair: From deinstitutionalization to homelessness.* Polity Press.

Del Casino, V., Jr. (2010) *Living with and experiencing disease.* In Brown, T., McLafferty, S., & Moon, G. (Eds.), *A companion to health and medical geography.* John Wiley & Sons.

DeLanda, M. (2006). *A new philosophy of society: Assemblage theory and social complexity.* A & C Black.

Deleuze, G. (1986). *Cinema one: The movement-image,* Continuum Books.

Deleuze, G. (1989). *Cinema two: The time-image,* Continuum Books.

Deleuze, G. (1988). *Spinoza: Practical philosophy.* City Lights Books.

Deleuze, G. (1992). Ethology: Spinoza and us. In Crary, J., & Kwinter, S. (Eds.), *Incorporations.* Zone Books, 625–633.

Deleuze, G. (1993). *The fold: Leibniz and the baroque.* University of Minnesota Press.

Deleuze G. (1988). *Spinoza: Practical philosophy.* City Lights Books.

Deleuze, G. (1995) *Negotiations.* Columbia University Press.

Deleuze, G., & Guattari, F. (1988). *A thousand plateaus: Capitalism and schizophrenia*. Bloomsbury Publishing.

Deleuze, G., & Guattari, F. (1994). *What is philosophy?* (H. Tomlinson & G. Burchell, Trans.). Columbia University Press.

Deleuze, G., & Guattari, F. (2003). *Mille piani*. Castelvecchi.

Deleuze, G., & Parnet, C. (2006). *Dialogues II*. Continuum.

DeVerteuil, G., & Andrews, G. (2007). Surviving profoundly unhealthy places: The ambivalent, fragile and absent therapeutic landscapes of the Soviet gulag. In Williams, A. (Ed.), *Therapeutic landscapes*. Ashgate.

DeVerteuil, G., & Wilton, R. (2009). Spaces of abeyance, care and survival: The addiction treatment system as a site of 'regulatory richness'. *Political Geography*, *28*(8), 463–472.

Dewsbury, J. D. (2000). Performativity and the event: Enacting a philosophy of difference. *Environment and Planning D: Society and Space*, *18*(4), 473–496.

Dewsbury, J. D. (2003). Witnessing space: 'Knowledge without contemplation'. *Environment and Planning A*, *35*(11), 1907–1932.

Dewsbury, J. D. (2009). Performative, non-representational and affect-based research: Seven injunctions. *The Sage handbook of qualitative geography*. Sage.

Dewsbury, J. D. (2011). The Deleuze-Guattarian assemblage: Plastic habits. *Area*, *43*(2), 148–153.

Dewsbury, J. D. (2012). Affective habit ecologies: Material dispositions and immanent inhabitations. *Performance Research*, *17*(4), 74–82.

Dewsbury, J. D. (2015). Non-representational landscapes and the performative affective forces of habit: From 'live' to 'blank'. *Cultural Geographies*, *22*(1), 29–47.

Dewsbury, J. D., & Bissell, D. (2015). Habit geographies: The perilous zones in the life of the individual. *Cultural Geographies*, *22*(1), 21–28.

Dewsbury, J. D., Harrison, P., Rose, M., & Wylie, J. (2002) Enacting geographies. *Geoforum 33*(4), 437–440.

Diez Roux, A. V. (2001). Investigating neighborhood and area effects on health. *American Journal of Public Health*, *91*(11), 1783–1789.

Dirksmeier, P., & Helbrecht, I. (2008). Time, non-representational theory and the 'performative turn': Towards a new methodology in qualitative social research. *Qualitative Social Research*, *9*(2).

Doel, M. A. (1996). A hundred thousand lines of flight: A machinic introduction to the nomad thought and scrumpled geography of Gilles Deleuze and Félix Guattari. *Environment and Planning D: Society and Space*, *14*(4), 421–439.

Doel, M. A., & Segrott, J. (2004). Materializing complementary and alternative medicine: Aromatherapy, chiropractic, and Chinese herbal medicine in the UK. *Geoforum*, *35*(6), 727–738.

Donovan, R., & Williams, A. (2007). Home as therapeutic landscape: Family caregivers providing palliative care at home. In Williams, A. (Ed.), *Therapeutic landscapes*. Ashgate, 199–218.

Dorn, M., & Laws, G. (1994). Social theory, body politics, and medical geography: Extending Kearns's invitation. *The Professional Geographer*, *46*(1), 106–110.

Doughty, K. (2013). Walking together: The embodied and mobile production of a therapeutic landscape. *Health & Place*, *24*, 140–146.

Ducey, A. (2007). More than a job: Meaning, affect, and training health care workers. In Clough, P., & Halley, J. (Eds.), *The affective turn: Theorizing the social*. Duke University Press, 187–208.

Ducey, A. (2010). Technologies of caring labor: From objects to affect. In Parrenas, R,. & Boris, E. (Eds.), *Intimate labors: Cultures, technologies, and the politics of care.* Stanford University Press, 18–32.

Duff, C. (2009). The drifting city: The role of affect and repair in the development of 'enabling environments'. *International Journal of Drug Policy*, *20*(3), 202–208.

Duff, C. (2010). Towards a developmental ethology: Exploring Deleuze's contribution to the study of health and human development. *Health*, *14*(6), 619–634.

Duff, C. (2011). Networks, resources and agencies: On the character and production of enabling places. *Health & Place*, *17*(1), 149–156.

Duff, C. (2012). Exploring the role of 'enabling places' in promoting recovery from mental illness: A qualitative test of a relational model. *Health & Place*, *18*(6) 1388–1395.

Duff, C. (2014a). *Assemblages of health: Deleuze's empiricism and the ethology of life.* Springer.

Duff, C. (2014b). The place and time of drugs. *International Journal of Drug Policy*, *25*(3), 633–639.

Duff, C. (2016a). Atmospheres of recovery: Assemblages of health. *Environment and Planning A*, *48*(1), 58–74.

Duff, C. (2016b). Assemblages, territories, contexts. *International Journal of Drug Policy*, *33*, 15–20.

Duff, C., & Moore, D. (2015). Going out, getting about: Atmospheres of mobility in Melbourne's night-time economy. *Social & Cultural Geography*, *16*(3), 299–314.

Duncan, C., Jones, K., & Moon, G. (1996). Health-related behaviour in context: A multilevel modelling approach. *Social Science & Medicine*, *42*(6), 817–830.

Duncan, C., Jones, K., & Moon, G. (1998). Context, composition and heterogeneity: Using multilevel models in health research. *Social Science & Medicine*, *46*(1), 97–117.

Dunn, J., R. (2000). Housing and health inequalities: Review and prospects for research. *Housing Studies*, *15*(3), 341–366.

Dunn, J. R., & Dyck, I. (2000). Social determinants of health in Canada's immigrant population: Results from the National Population Health Survey. *Social Science & Medicine*, *51*(11), 1573–1593.

Durie, R., & Wyatt, K. (2007). New communities, new relations: the impact of community organization on health outcomes. *Social Science & Medicine*, *65*(9), 1928–1941.

Dyck, I. (1995). Hidden geographies: The changing lifeworlds of women with multiple sclerosis. *Social Science & Medicine*, *40*(3), 307–320.

Dyck, I., Kontos, P., Angus, J., McKeever, P., & Poland, B. (2005). The home as a site of long-term care: Meanings and management of bodies and spaces. *Health & Place*, *11*(2) 173–185.

Ecob, R., & Macintyre, S. (2000). Small area variations in health related behaviours: Do these depend on the behaviour itself, its measurement, or on personal characteristics? *Health & Place*, *6*(4), 261–274.

Edensor, T., & Richards, S. (2007). Snowboarders vs skiers: Contested choreographies of the slopes. *Leisure Studies*, *26*(1), 97–114.

Edwards, J. B. (1996). Weather-related road accidents in England and Wales: A spatial analysis. *Journal of Transport Geography*, *4*(3), 201–212.

Ellaway, A., & Macintyre, S. (2009) Neighborhoods and health. In Brown, T., McLafferty, S., & Moon, G. (Eds.), *A companion to health and medical geography.* Wiley-Blackwell, 399–417.

Ellaway, A., Kirk, A., Macintyre, S., & Mutrie, N. (2007). Nowhere to play? The relationship between the location of outdoor play areas and deprivation in Glasgow. *Health & Place, 13*, 557–561.

English, J., Wilson, K., & Keller-Olaman, S. (2008). Health, healing and recovery: Therapeutic landscapes and the everyday lives of breast cancer survivors. *Social Science & Medicine, 67*(1), 68–78.

Evans B. (2010). Anticipating fatness: Childhood, affect and the pre-emptive 'war on obesity'. *Transactions of the Institute of British Geographers, 35*(1), 21–38.

Evans, J. D. (2011). Exploring the (bio)political dimensions of voluntarism and care in the city: The case of a 'low barrier' emergency shelter. *Health & Place, 17*, 24–3.

Evans, J. (2012). Supportive measures, enabling restraint: Governing homeless street drinkers in Hamilton, Canada. *Social & Cultural Geography, 13*(2), 175–190.

Evans, J. D. (2014). Painting therapeutic landscapes with sound: On land by Brian Eno. In Kearns, R., & Andrews, G. J. (Eds.). *Soundscapes of wellbeing in popular music.* Ashgate.

Evans, J. D., Crooks, V. A., & Kingsbury, P. T. (2009). Theoretical injections: On the therapeutic aesthetics of medical spaces. *Social Science & Medicine, 69*(5), 716–721.

Evers, C. (2009). 'The point': Surfing, geography and a sensual life of men and masculinity on the Gold Coast, Australia. *Social & Cultural Geography, 10*(8), 893–908.

Ewing, R., & Handy, S. (2009). Measuring the unmeasurable: Urban design qualities related to walkability. *Journal of Urban Design, 14*(1), 65–84.

Eyles, J. (1985). *Senses of place.* Silverbook Press.

Eyles, J. (1990). How significant are the spatial configurations of health care systems? *Social Science & Medicine, 30*(1), 157–164.

Eyles, J., & Donovan, J. (1986). Making sense of sickness and care: An ethnography of health in a West Midlands town. *Transactions of the Institute of British Geographers*, 415–427.

Farmer, J., Lauder, W., Richards, H., & Sharkey, S. (2003). Dr. John has gone: Assessing health professionals' contribution to remote rural community sustainability in the UK. *Social Science & Medicine, 57*(4), 673–686.

Fleuret, S., & Atkinson, S. (2007). Wellbeing, health and geography: A critical review and research agenda. *New Zealand Geographer, 63*(2), 106–118.

Fleuret, S., & Prugneau, J. (2015). Assessing students' wellbeing in a spatial dimension. *The Geographical Journal, 181*, 110–120.

Foley, R. (2010). *Healing waters: Therapeutic landscapes in historic and contemporary Ireland.* Ashgate.

Foley, R. (2011). Performing health in place: The holy well as a therapeutic assemblage. *Health & Place, 17*(2), 470–479.

Foley, R. (2014). The Roman-Irish bath: Medical/health history as therapeutic assemblage. *Social Science & Medicine, 106*, 10–19.

Foley, R. (2015). Swimming in Ireland: Immersions in therapeutic blue space. *Health & Place, 35*(5), 218–225.

Foley, R., & Kistemann, T. (2015). Blue space geographies: Enabling health in place. *Health & Place, 35*, 157–165.

Foley, R., Wheeler, A., & Kearns, R. (2011). Selling the colonial spa town: The contested therapeutic landscapes of Lisdoonvarna and Te Aroha. *Irish Geography, 44*(2–3), 151–172.

Foo, K., Martin, D., Polsky, C., Wool, C., & Ziemer, M (2015). Social well-being and environmental governance in urban neighbourhoods in Boston, MA. *The Geographical Journal, 181*(2), 138–146.

Ford, R. G., & Smith, G. C. (2008). Geographical and structural change in nursing care provision for older people in England, 1993–2001. *Geoforum, 39*(1), 483–498.

Foucault, M. (1997). *Ethics*. The New Press.

Fox, N. J. (2011). The ill-health assemblage: Beyond the body-with-organs. *Health Sociology Review, 20*(4), 359–371.

Fraser, L. K., & Edwards, K. L. (2010). The association between the geography of fast food outlets and childhood obesity rates in Leeds, UK. *Health & Place, 16*, 1124–1128.

Gagnon, M., & Holmes, D. (2016). Body–drug assemblages: Theorizing the experience of side effects in the context of HIV treatment. *Nursing Philosophy, 17*(4), 250–261.

Gallagher, M., & Prior, J. (2014). Sonic geographies: Exploring phonographic methods. Progress in Human Geography, *38*(2), 267–284.

Gant, R. (1997). Elderly people, personal mobility and local environment. *Geography*, July, 207–217.

Garfield, L. (2012). Dreaming of infrastructure: Architexture as control and Parkour as rebellion. *Intertext, 20*(1), 17.

Garhammer, M. (2002). Pace of life and enjoyment of life. *Journal of Happiness Studies, 3*(3), 217–256.

Gastaldo, D., Andrews, G. J., & Khanlou, N. (2004). Therapeutic landscapes of the mind: Theorizing some intersections between health geography, health promotion and immigration studies. *Critical Public Health, 14*(2), 157–176.

Gatrell, A. C. (2005). Complexity theory and geographies of health: A critical assessment. *Social Science & Medicine, 60*(12), 2661–2671.

Gatrell, A. C. (2011). *Mobilities and health*. Ashgate.

Gatrell, A. C. (2013). Therapeutic mobilities: Walking and 'steps' to wellbeing and health. *Health & Place, 22*, 98–106.

Gehl, J. (2010). *Cities for people*. Island Press.

George, S. (2015). 'Real nursing work' versus 'charting and sweet talking': The challenges of incorporating into US urban health care settings for Indian immigrant nurses. In Parry, B., Greenhough, B., Brown, T., & Dyck, I. (Eds.), *Bodies across borders: The global circulation of body parts, medical tourists and professionals*. Ashgate.

Gesler, W. M. (1992a). *The cultural geography of health care*. University of Pittsburgh Press.

Gesler, W. M. (1992b). Therapeutic landscapes: Medical issues in light of the new cultural geography. *Social Science & Medicine, 34*(7), 735–746.

Gesler, W. (1996). Lourdes: Healing in a place of pilgrimage. *Health & Place, 2*(2), 95–105.

Gesler, W. M. (1999). Words in wards: language, health and place. Health & Place, *5*(1), 13–25.

Gesler, W. (2000). Hans Castorp's journey-to-knowledge of disease and health in Thomas Mann's *The magic mountain. Health & Place, 6*(2), 125–134.

Gesler, W. M. (2003) *Therapeutic hospital environments*. In Gesler, W. M. (Ed.), *Healing places*. Rowman and Littlefield.

Gesler, W., & Curtis, S. (2007). Application of concepts of therapeutic landscapes to the design of hospitals in the UK: The example of a mental health facility in London. In Williams , A. (Ed.), *Therapeutic landscapes*. Ashgate, 149–164.

Gesler, W. M., Kearns, R. A. (2002). *Culture/place/health*. Routledge.

Gesler, W. M., Bell, M., Hubbard. P., & Francis, S. (2004). Therapy by design: Evaluating the UK hospital building program. *Health & Place, 10*(2), 117–128.

Giles-Corti, B., Wood, G., Pikora, T., Learnihan, V., Bulsara, M., van Niel, K., ... Villanueva, K. (2011). School site and the potential to walk to school: The impact of street connectivity and traffic exposure in school neighbourhoods. *Health & Place, 17*, 545–550.

Gilmour, J. A. (2006). Hybrid space: Constituting the hospital as a home space for patients. *Nursing Inquiry, 13*(1), 16–22.

Gilroy, R. (2012). *Wellbeing and the neighbourhood: Promoting choice and independence for all ages*. In Atkinson, S., Fuller, S., & Painter, J. (Eds.), *Wellbeing and place*. Ashgate.

Gleeson, B. (1999). *Geographies of disability*. Psychology Press.

Glennie, P., & Thrift, N. (2009). *Shaping the day: A history of timekeeping in England and Wales 1300–1800*. Oxford University Press.

Golledge, R. G. (1993). Geography and the disabled: A survey with special reference to vision impaired and blind populations. *Transactions of the Institute of British Geographers*, 63–85.

Gorman, R. (2017). Smelling therapeutic landscapes: Embodied encounters within spaces of care farming. *Health & Place, 47*, 22–28.

Gould, P. (1993). *The slow plague: A geography of the AIDS pandemic*. Blackwell.

Gould, M. (2010). Modelling chronic disease. In Brown, T., McLafferty, S., & Moon, G. (Eds.), *Companion to health and medical geography*. Wiley-Blackwell.

Green, K. (2011). It hurts so it is real: Sensing the seduction of mixed martial arts. *Social & Cultural Geography, 12*,(4), 377–396.

Greenhough, B. (2006). Decontextualised? dissociated? detached? Mapping the networks of bio-informatic exchange. *Environment and Planning A, 38*(3), 445–463.

Greenhough, B. (2010). Vitalist geographies: Life and the more-than-human. Anderson, B., & Harrison, P. (Eds.), *Taking place: Non-representational theories and geography*. Ashgate.

Greenhough, B. (2011a). Assembling an island laboratory. *Area, 43*(2), 134–138.

Greenhough, B. (2011b). Citizenship, care and companionship: Approaching geographies of health and bioscience. *Progress in Human Geography, 35*(2), 153–171.

Greenhough, B. (2012). Where species meet and mingle: Endemic human-virus relations, embodied communication and more-than-human agency at the Common Cold Unit 1946–90. *Cultural Geographies, 19*(3), 281–301.

Greenhough, B., & Roe, E. J. (2011). Ethics, space, and somatic sensibilities: Comparing relationships between scientific researchers and their human and animal experimental subjects. *Environment and Planning D: Society and Space, 29*(1), 47–66.

Halford, S., & Leonard, P. (2003). Space and place in the construction and performance of gendered nursing identities. *Journal of Advanced Nursing, 42*(2), 201–208.

Hall, E. (2000). 'Blood, brains and bones': Taking the body seriously in the geography of health and impairment. *Area, 32*(1), 21–30.

Hall, E. (2003). Reading maps of the genes: Interpreting the spatiality of genetic knowledge. *Health & Place 9*, 151–161.

Hall, E. (2004). Spaces and networks of genetic knowledge making: The 'geneticisation' of heart disease. *Health & Place, 10*(4) 311–318.

Hall, E. & Wilton, R. (2017). Towards a relational geography of disability. *Progress in Human Geography* (in press).

Hanlon, N. T. (2001). Sense of place, organizational context and the strategic management of publicly funded hospitals. *Health Policy*, *58*(2), 151–173.

Hanlon, N. (2014). Doing health geography with feeling. *Social Science & Medicine*, *115*, 144–146.

Hanlon, N., Skinner, M. W., Joseph, A. E., Ryser, L., & Halseth, G., (2014). Place integration through efforts to support healthy aging in resource frontier communities: The role of voluntary sector leadership. *Health & Place*, *29*, 132–139.

Harvey, D. (1990). Between space and time: Reflections on the geographical imagination. *Annals of the Association of American Geographers*, *80*(3), 418–434.

Hay, G., Whigham, P., Kypri, K., & Langley, J. (2009). Neighborhood deprivation and access to alcohol outlets: A national study. *Health & Place*, *15*, 1086–1093.

Hinchliffe, S., & Ward, K. J. (2014). Geographies of folded life: How immunity reframes biosecurity. *Geoforum*, 53, 136–144.

Hinchliffe, S., Bingham, N., Allen, J., & Carter, S. (2016). *Pathological lives: disease, space and biopolitics*. John Wiley & Sons.

Hodgins, H. J., & Wuest, J. (2007). Uncovering factors affecting use of the emergency department for less urgent health problems in urban and rural areas. *Canadian Journal of Nursing Research 39*(3), 78–102.

Hoyez, A. C. (2007a). The 'world of yoga': The production and reproduction of therapeutic landscapes. *Social Science & Medicine*, *65*(1), 112–124.

Hoyez, A., C. (2007b). From Rishikesh to Yogaville: The globalization of therapeutic landscapes. In Williams, A. (Ed.), *Therapeutic landscapes*. Ashgate.

Hubbard, P., & Lilley, K. (2004). Pacemaking the modern city: The urban politics of speed and slowness. *Environment and Planning D: Society and Space*, *22*(2), 273–294.

Humberstone, B. (2011). Embodiment and social and environmental action in nature-based sport: Spiritual spaces. *Leisure Studies*, *30*(4), 495–512.

Ingold, T. (2015). Foreword. In Vannini, P. (Ed.), *Non-representational methodologies: Re-envisioning research*. Routledge.

Iredale, R., Jones, L., Gray, J., & Deaville, J. (2005). 'The edge effect': an exploratory study of some factors affecting referrals to cancer genetic services in rural Wales. *Health & Place*, *11*(3), 197–204.

Jackson, P., & Neely, A. H. (2015). Triangulating health: Toward a practice of a political ecology of health. *Progress in Human Geography*, *39*(1), 47–64.

Jacobs, J., & Nash, C. (2003). Too little, too much: Cultural feminist geographies. *Gender, Place and Culture 10*(3): 265–279.

Jarvis, H., Pain, R., Pooley, C. (2011). Multiple scales of time-space and lifecourse. *Environment and Planning A*, *43*(3), 519–524.

Jayne, M., Valentine, G., & Holloway, S. L. (2010). Emotional, embodied and affective geographies of alcohol, drinking and drunkenness. *Transactions of the Institute of British Geographers*, *35*(4), 540–554.

Jerrett, M., Burnett, R. T., Ma, R., Pope, C. A., III, Krewski, D., Newbold, K. B., & Thun, M. J. (2005). Spatial analysis of air pollution and mortality in Los Angeles. *Epidemiology*, *16*(6), 727–736.

Jones, O. (2008). Stepping from the wreckage: Geography, pragmatism and anti-representational theory. *Geoforum*, *39*(4), 1600–1612.

Jones, M. (2009). Phase space: Geography, relational thinking, and beyond. *Progress in Human Geography, 33*(4), 487–506.

Joseph, A. E., & Chalmers, A. L. (1996). Restructuring long-term care and the geography of ageing: A view from rural New Zealand. *Social Science & Medicine, 42*(6), 887–896.

Joseph, A. E., & Cloutier , D. S. (1991). Elderly migration and its implications for service provision in rural communities: An Ontario perspective. *Journal of Rural Studies, 7*(4), 433–444.

Joseph, A. E., & Hallman, B. C. (1998). Over the hill and far away: Distance as a barrier to the provision of assistance to elderly relatives. *Social Science & Medicine, 46*(6), 631–639.

Joseph, A. E., & Kearns, R. A. (1996). Deinstitutionalization meets restructuring: The closure of a psychiatric hospital in New Zealand. *Health & Place, 2*(3), 179–189.

Joseph, A. E., & Phillips, D. R. (1984). *Accessibility and utilization: Geographical perspectives on health care delivery.* Sage.

Joseph, A. E., Kearns, R. A., & Moon, G. (2009). Recycling former psychiatric hospitals in New Zealand: Echoes of deinstitutionalisation and restructuring. *Health & Place, 15*, 79–87.

Justesen, L., Gyimóthy, S., & Mikkelsen, B. E. (2014). Moments of hospitality: Rethinking hospital meals through a non-representational approach. *Hospitality & Society, 4*(3), 231–248.

Kearns, R. A. (1993). Place and health: Towards a reformed medical geography. *The Professional Geographer, 45*(2), 139–147.

Kearns, R. A. (1994a). Putting health and health care into place: An invitation accepted and declined. *The Professional Geographer, 46*(1), 111–115.

Kearns, R. A. (1994b). To reform is not to discard: A reply to Paul. *The Professional Geographer, 46*(4), 505–507.

Kearns, R. A. (2014). The health in 'life's infinite doings': A response to Andrews et al. *Social Science & Medicine 115*(1), 147–149.

Kearns, R. A., & Andrews, G. J. (2010). Geographies of wellbeing. In Smith, S. J., Pain, R., Marston, S. A., & Jones J. P. P., III (Eds.), *The Sage handbook of social geographies.* Sage.

Kearns, R. A., & Barnett, R. (1992). Enter the supermarket: Entrepreneurial medical practice in New Zealand. *Environment and Planning C: Government and Policy, 10*(3), 267–281.

Kearns, R. A., & Barnett, R. (1997). Consumerist ideology and the symbolic landscapes of private medicine. *Health & Place, 3*(3), 171–180.

Kearns, R. A., & Barnett, R. (1999). To boldly go? Place, metaphor and the marketing of Auckland's Starship Hospital. *Environment and Planning D: Society and Space 17*(2), 201–226.

Kearns, R. A., & Barnett, R. (2000). 'Happy Meals' in the Starship Enterprise: Interpreting a moral geography of health care consumption. *Health & Place, 6*, 81–93.

Kearns, R. A., & Collins, D. C. (2000). New Zealand children's health camps: Therapeutic landscapes meet the contract state. *Social Science & Medicine, 51*(7), 1047–1059.

Kearns, R. A., Barnett, R., & Newman, D. (2003). Placing private health care: Reading Ascot hospital in the landscape of contemporary Auckland. *Social Science & Medicine, 56*(11), 2303–2315.

Kearns, R., Joseph, A., & Moon, G. (2015). *The afterlives of the psychiatric asylum: The recycling of concepts, sites and memories*. Ashgate.

Keil, R., & Ali, H. (2006). Multiculturalism, racism and infectious disease in the global city: The experience of the 2003 SARS outbreak in Toronto. *Topia: Canadian Journal of Cultural Studies*, (16), 23–34.

Keil, R., & Ali, H. (2007). Governing the sick city: Urban governance in the age of emerging infectious disease. *Antipode*, *39*(5), 846–873.

Kelly, M. P., & Field, D. (1996). Medical sociology, chronic illness and the body. *Sociology of Health & Illness*, *18*(2), 241–257.

King, B. (2010). Political ecologies of health. *Progress in Human Geography*, *34*(1), 38–55.

Kingma, M. (2006). *Nurses on the move: Migration and the global health care economy*. Cornell University Press.

Kingsbury, P., Crooks, V. A., Snyder, J., Johnston, R., & Adams, K. (2012). Narratives of emotion and anxiety in medical tourism: On *State of the heart* and *Larry's kidney*. *Social & Cultural Geography*, *13*(4), 361–378.

Kong, L., Yeoh, B., & Teo, P. (1996). Singapore and the experience of place in old age. *Geographical Review*, 529–549.

Kraftl, P. (2013). *Geographies of alternative education*. Policy Press.

Kraftl, P., & Horton, J. (2007) 'The health event': Everyday, affective politics of participation. *Geoforum*, *38*(5), 1012–1027.

Kwan, M. P. (2012). The uncertain geographic context problem. *Annals of the Association of American Geographers*, *102*(5), 958–968.

Lacan, J. (1992). *The seminar of Jacques Lacan book VII: The ethics of psychoanalysis, 1959–1960*. Norton.

Laditka, J. N. (2004). Physician supply, physician diversity, and outcomes of primary health care for older persons in the United States. *Health & Place*, *10*(3), 231–244.

Last, A. (2012). Experimental geographies. *Geography Compass*, *6*(12), 706–724.

Latham, A. (2003). Research, performance, and doing human geography: some reflections on the diary-photograph, diary-interview method. *Environment and Planning A*, *35*(11), 1993–2018.

Latham, A. M., D, P. (2008). Speed and slowness. In Hall, T., Hubbard, P., Rennie-Short, J. (Eds.), *The Sage Companion to the City*. Sage.

Lawler, L., & Leonard, V. (2016). Henri Bergson. Stanford Encyclopedia of Philosophy https://plato.stanford.edu/entries/bergson/.

Laws, J. (2009). Reworking therapeutic landscapes: The spatiality of an 'alternative'self-help group. *Social Science & Medicine*, *69*(12), 1827–1833.

Lea, J. (2008). Retreating to nature: Rethinking 'therapeutic landscapes'. *Area*, *40*(1), 90–98.

Lea J., Cadman L., & Philo C. (2015). Changing the habits of a lifetime? Mindfulness meditation and habitual geographies. *Cultural Geographies*, *22*(1), 49–65.

Lefebvre, H. (2004). *Rhythmanalysis: Space, time and everyday life*. Continuum.

Lehoux, P., Daudelin, G., Poland, B., Andrews, G. J., & Holmes, D. (2007). Designing a better place for patients: Professional struggles surrounding satellite and mobile dialysis units. *Social Science & Medicine*, *65*(7), 1536–1548.

Lee, A. C., & Maheswaran, R. (2011). The health benefits of urban green spaces: A review of the evidence. *Journal of Public Health*, *33*(2), 212–222.

Lee, V. S. P., Simpson, J., & Froggatt, K. (2013). A narrative exploration of older people's transitions into residential care. *Aging and Mental Health*, *17*(1), 48–56.

Lefebvre, H., (1992). *Elements de rhythmanalyse: Introduction a la connaissance des rhythms*. Editions Syllepse.

Leong, A. M., Crighton, E. J., Moineddin, R., Mamdani, M., & Upshur, R. E., (2006). Time series analysis of age related cataract hospitalizations and phacoemulsification. *BMC Ophthalmology*, *6*(1), 2.

Levine, R. (2005). A geography of busyness. *Social Research*, *72*(2), 355–370.

Levine, R. N. (2008). *A geography of time: On tempo, culture, and the pace of life*. Basic Books.

Liaschenko, J. (1997). Ethics and the geography of the nurse– patient relationship: Spatial vulnerable and gendered space. *Scholarly Inquiry for Nursing Practice* *11*(1), 45–59.

Liaschenko, L., Peden-McAlpine, C., & Andrews G. J. (2011). Institutional geographies in dying: Nurses actions and observations on dying spaces inside and outside intensive care units. *Health & Place*, *17*(3), 814–821.

Lin, L., & Moudon, A. V. (2010). Objective versus subjective measures of the built environment, which are most effective in capturing associations with walking? *Health & Place*, *16*, 339–348.

Lingis, A. (2015). Irrevocable loss. In Vannini, P. (Ed.), *Non-representational methodologies: Re-envisioning research*. Routledge.

Little, J. (2013). Pampering, well-being and women's bodies in the therapeutic spaces of the spa. *Social & Cultural Geography*, *14*(1), 41–58.

Little, D. (2016) Assemblage theory. Blog entry 15 November 2012. http://understandingsociety.blogspot.ca/2012/11/assemblage-theory.html.

Longhurst, R. (2000). Corporeographies of pregnancy: Bikini babes. *Environment and Planning D: Society and Space*, *18*(4), 453–472.

Longhurst, R. (2005). (Ad) dressing pregnant bodies in New Zealand: Clothing, fashion, subjectivities and spatialities. *Gender, Place & Culture*, *12*(4), 433–446.

Lorimer, H. (2005). Cultural geography: The busyness of being 'more-than-representational'. *Progress in Human Geography*, *29*(1), 83–94.

Lorimer, H. (2007). Cultural geography: Worldly shapes, differently arranged. *Progress in Human Geography*, *31*(1), 89–100.

Lorimer, H. (2008). Cultural geography: Nonrepresentational conditions and concerns. *Progress in Human Geography*, *32*(4), 551–559.

Lorimer, H. (2012). Surfaces and slopes. *Performance Research*, *17*, 83–86.

Lorimer, H. (2015). Crafting cultural theory at the Gates of Heaven. Lecture given at the VIII Annual Conference of the CECT. Video recording available at www.uttv. ee/naita?id=21932.

Loukaitou-Sideris, A. (2006). Is it safe to walk? Neighborhood safety and security considerations and their effects on walking. *Journal of Planning Literature*, *20*(3), 219–232.

Love, M., Wilton, R., & DeVerteuil, G. (2012). 'You have to make a new way of life': Women's drug treatment programmes as therapeutic landscapes in Canada. *Gender, Place & Culture*, *19*(3), 382–396.

Luginaah, I. (2008). Local gin (akpeteshie) and HIV/AIDS in the Upper West Region of Ghana: The need for preventive health policy. *Health & Place*, *14*(4), 806–816.

Maas, J., Verheij, R. A., Groenewegen, P. P., De Vries, S., & Spreeuwenberg, P. (2006). Green space, urbanity, and health: How strong is the relation? *Journal of Epidemiology & Community Health*, *60*(7), 587–592.

Maas, B., Fairbairn, N., Kerr, T., Li, K., Montaner, J., & Wood, E. (2007). Neighborhood and HIV infection among IDU: Place of residence independently predicts HIV infection among a cohort of injection drug users. *Health & Place*, *13*, 432–439.

Macintyre, S., MacIver, S., & Sooman, A. (1993). Area, class and health: Should we be focusing on places or people? *Journal of Social Policy*, *22*(2), 213–234.

Macintyre, S., Ellaway, A., & Cummins, S. (2002). Place effects on health: How can we conceptualise, operationalise and measure them? *Social Science & Medicine*, *55*(1), 125–139.

MacKian, S. (2000). Contours of coping: Mapping the subject world of long-term illness. *Health & Place*, *6*(2), 95–104.

MacKian, S. C. (2008). What the papers say: Reading therapeutic landscapes of women's health and empowerment in Uganda. *Health & Place*, *14*(1), 106–115.

MacKian, S. C. (2009). Wellbeing. In Kitchin, R., & Thrift, N. (Eds.), *International encyclopedia of human geography*. Elsevier, 235–240.

Macpherson, H. (2008). 'I don't know why they call it the Lake District they might as well call it the rock district!' The workings of humour and laughter in research with members of visually impaired walking groups. *Environment and Planning D: Society and Space*, *26*(6), 1080–1095.

Macpherson, H. (2009a). The intercorporeal emergence of landscape: Negotiating sight, blindness, and ideas of landscape in the British countryside. *Environment and Planning A*, *41*(5), 1042.

Macpherson, H. (2009b). Articulating blind touch: Thinking through the feet. *The Senses and Society*, *4*(2), 179–193.

Magilvy, J. K., & Congdon, J. G. (2000). The crisis nature of health care transitions for rural older adults. *Public Health Nursing*, *17*(5), 336–345.

Mahon-Daly, P., & Andrews, G. J. (2002). Liminality and breastfeeding: Women negotiating space and two bodies. *Health & Place*, *8*(2), 61–76.

Malins, P. (2004). Body–space assemblages and folds: Theorizing the relationship between injecting drug user bodies and urban space. *Continuum: Journal of Media & Cultural Studies*, *18*(4), 483–495.

Malone, R. (2003). Distal nursing. *Social Science & Medicine,* *56*(11), 2317–2326.

Malpas, J. (2012). Putting space in place: Philosophical topography and relational geography. *Environment and Planning D: Society and Space*, *30*(2), 226–242.

Mansvelt, J. (1997). Working at leisure: Critical geographies of ageing. *Area*, *29*(4), 289–298.

Marcus, G. E., & Saka, E. (2006). Assemblage. *Theory, Culture & Society*, *23*(2–3), 101–106.

Marston, S. A., Jones, J. P., & Woodward, K. (2005). Human geography without scale. *Transactions of the Institute of British Geographers*, *30*(4), 416–432.

Massey, D. (1992). Politics and space/time. *New Left Review*, 65–84.

Massumi, B. (2002). *Parables for the virtual: Movement, affect, sensation*. Duke University Press.

Matthews, S. A., Moudon, A. V., & Daniel, M. (2009). Work group II: Using geographic information systems for enhancing research relevant to policy on diet, physical activity, and weight. *American Journal of Preventive Medicine*, *36*(4), 171–176.

Mattson, J. W. (2010). Aging and mobility in rural and small urban areas: A survey of North Dakota. *Journal of Applied Gerontology, 30*(6), 700–718.

May, J. M. (1959). The ecology of human disease. *Studies in medical geography: Vol.1.* MD Publications.

May, J., & Thrift, N. (Eds.). (2003). *Timespace: geographies of temporality.* Routledge.

Mayer, J. (2010). Medical geography. In Brown, T., McLafferty, S., & Moon, G. (Eds.), *A companion to health and medical geography*. John Wiley & Sons.

Mayer, J. D., & Meade M. S. (1994). A reformed medical geography reconsidered. *The Professional Geographer, 46*(1), 103–106.

McCann, E. (2011). Veritable inventions: Cities, policies and assemblage. *Area, 43*(2), 143–147.

McCormack, D. P. (2002). A paper with an interest in rhythm. *Geoforum, 33*(4), 469–485.

McCormack, D. P. (2003). An event of geographical ethics in spaces of affect. *Transactions of the Institute of British Geographers, 28*(4), 488–507.

McCormack, D. P. (2007). Molecular affects in human geographies. *Environment and Planning A, 39*(2), 359.

McCormack, D. P. (2008). Geographies for moving bodies: Thinking, dancing, spaces. *Geography Compass, 2*(6), 1822–1836.

McCormack, D. P. (2013). *Refrains for moving bodies: Experience and experiment in affective spaces.* Duke University Press.

McGrath, L., & Reavey, P. (2015). Seeking fluid possibility and solid ground: Space and movement in mental health service users' experiences of 'crisis'. *Social Science & Medicine, 128,* 115–125.

McGrath, L., & Reavey, P. (2016). 'Zip me up, and cool me down': Molar narratives and molecular intensities in 'helicopter' mental health services. *Health & Place, 38,* 61–69.

McHugh, K. E. (2009). Movement, memory, landscape: An excursion in non-representational thought. *GeoJournal, 74*: 209–218.

McLafferty, S. L. (2003). GIS and health care. *Annual Review of Public Health, 24*(1), 25–42.

Meade, M. S. (1977). Medical geography as human ecology: The dimension of population movement. *Geographical Review, 67*(4) 379–393.

Meade, M. S., & Earickson, R. (2000). Medical geography. Guilford Press.

Meade, M. S., & Emch, M. (2010). Medical geography. Guilford Press.

Mercado, R., Páez, A., & Newbold, K. B. (2010). Transport policy and the provision of mobility options in an aging society: A case study of Ontario, Canada. *Journal of Transport Geography, 18*(5), 649–661.

Merleau-Ponty, M. (1962). *Phenomenology of perception.* (C. V. Smith, Trans.). Routledge and Kegan Paul.

Merriman, P. (2012). Human geography without time-space. *Transactions of the Institute of British Geographers, 37*(1), 13–27.

Merriman, P. (2013). Unpicking space-time: Towards new apprehensions of movement-space. In Ehland, C. (Ed.), *Space and mobility*. Rodopi, 177–192.

Merriman, P. (2016). Mobility infrastructures: Modern visions, affective environments and the problem of car parking. *Mobilities, 11,* 83–98.

Merriman, P., Jones, M., Olsson, G., Sheppard, E., Thrift, N., & Tuan, Y. F. (2012). Space and spatiality in theory. *Dialogues in Human Geography, 2*(1), 3–22.

Meyer J W, Speare, A., Jr. (1985). Distinctively elderly mobility: Types and determinants. *Economic Geography*, *61*(1), 79–88.

Middleton, J. (2009). 'Stepping in time': Walking, time and space in the city. *Environment and Planning* A, *41*, 1943–1961.

Middleton, J., (2010). Sense and the city: Exploring the embodied geographies of urban walking. *Social & Cultural Geography*, *11*(6), 575–596.

Middleton, J. (2011). 'I'm on autopilot, I just follow the route': Exploring the habits, routines, and decision-making practices of everyday urban mobilities. *Environment and Planning A*, *43*, 2857–2877.

Miles, R. (2006). Neighborhood disorder and smoking: Findings of a European urban survey. *Social Science & Medicine*, *63*(9), 2464–2475.

Miller, H. J. (2005). A measurement theory for time geography. *Geographical Analysis*, *37*(1), 17–45.

Milligan, C. (2000). Bearing the burden: Towards a restructured geography of caring. *Area*, *32*(1), 49–58.

Milligan, C. (2009). *There's no place like home: place and care in an aging society*. Ashgate.

Milligan, C., & Bingley, A. (2007). Restorative places or scary spaces? The impact of woodland on the mental well-being of young adults. *Health & Place*, *13*(4), 799–811.

Milligan, C., & Wiles, J. (2010). Landscapes of care. *Progress in Human Geography*, *34*(6), 736–754.

Milligan, C., Gatrell, A., & Bingley, A. (2004). 'Cultivating health': Therapeutic landscapes and older people in northern England. *Social Science & Medicine*, *58*(9), 1781–1793.

Mohan, J. F. (1998). Explaining geographies of health care: A critique. *Health & Place*, *4*(2), 113–124.

Mohan, J., Twigg, L., Barnard, S., & Jones, K. (2005). Social capital, geography and health: A small-area analysis for England. *Social Science & Medicine*, *60*(6), 1267–1283.

Moon, G., & Brown, T. (1998). *Place, space and the reform of the British NHS*. In Kearns, R. A., & Gesler, W. (Eds.), *Putting health into place*. Syracuse University Press.

Moon, G., & Brown, T. (2000). Governmentality and the spatialized discourse of policy: The consolidation of the Post-1989 NHS reforms. *Transactions of the Institute of British Geographers*, *25*(1), 65–76.

Moon, G., & Brown, T. (2001). Closing Barts: Community and resistance in contemporary UK hospital policy. *Environment and Planning D: Society and Space 19*, 43–59.

Moon, G., & North, N. (2000). *Policy and place: General medical practice in the UK*. Macmillan.

Moon, G., Mohan, J., Twigg, L., McGrath, K., & Pollock, A. (2002). Catching waves: The historical geography of the general practice fundholding initiative in England and Wales. *Social Science & Medicine*, *55*(12), 2201–2213.

Moon, G., Joseph, A. E., & Kearns, R. A. (2005). Towards a general explanation for the survival of the private asylum. *Environment and Planning C: Government and Policy*, *23*, 159–172.

Moon, G., Kearns, R. A., & Joseph, A. E. (2006). Selling the private asylum: Therapeutic landscapes and the (re)valorization of confinement in the era of community care. *Transactions Institute British Geographers*, *NS 31*, 131–149.

Moore, E. G., & Pacey, M. A. (2004). Geographic dimensions of aging in Canada, 1991–2001. *Canadian Journal on Aging*, *23*(5), S5–S21.

Moss, P. (1997). Negotiating spaces in home environments: Older women living with arthritis. *Social Science & Medicine, 45*(1), 23–33.

Moss, P., & Dyck, I. (1999). Body, corporeal space, and legitimating chronic illness: Women diagnosed with ME. *Antipode, 31*(4), 372–397.

Moss, P., & Dyck, I. (2003). *Women, body, illness: Space and identity in the everyday lives of women with chronic illness.* Rowman & Littlefield.

Muirhead, S. (2012). Exploring embodied and emotional experiences within the landscapes of environmental volunteering. In Atkinson, S., Fuller, S., & Painter, J. (Eds.), *Wellbeing and place.* Ashgate.

Nash, M. (2012). 'Working out' for two: Performances of 'fitness' and femininity in Australian prenatal aerobics classes. *Gender, Place & Culture, 19*(4), 449–471.

Nemet, G. F., & Bailey, A. J. (2000). Distance and health care utilization among the rural elderly. *Social Science & Medicine, 50*(9), 1197–1208.

Norris, P. (1997). The state and the market: The impact of pharmacy licensing on the geographical distribution of pharmacies. *Health & Place, 3*(4), 259–269.

Oliver, C. (2007). *Retirement migration: Paradoxes of ageing.* Routledge.

Oppong, J. R., & Harold, A. (2010). Disease, ecology and environment. In Brown, T., McLafferty, S., & Moon, G. (Eds.), *Companion to health and medical geography.* Wiley-Blackwell.

Pain, R. (2001). Gender, race, age and fear in the city. *Urban Studies, 38*(5–6), 899–913.

Pain R. (2006). Paranoid parenting? Rematerializing risk and fear for children. *Social & Cultural Geography, 7*(2), 221–243.

Palka, E. (1999). Accessible wilderness as a therapeutic landscape: experiencing the nature of Denali National Park, Alaska. In Williams, A. (Ed.), *Therapeutic landscapes: The dynamic between place and wellness.* Ashgate, 29–51.

Parr, H. (1997). Mental health, public space, and the city: Questions of individual and collective access. *Environment and Planning D: Society and Space, 15*(4), 435–454.

Parr H. (1998) Mental health, ethnography and the body. *Area, 30*(1), 28–37.

Parr, H. (2000). Interpreting the 'hidden social geographies' of mental health: Ethnographies of inclusion and exclusion in semi-institutional places. *Health & Place, 6*(3), 225–237.

Parr, H. (2002). Medical geography: Diagnosing the body in medical and health geography, 1999–2000. *Progress in Human Geography, 26*(2), 240–251.

Parr, H. (2003). Medical geography: Care and caring. *Progress in Human Geography, 27*(2), 212–221.

Parr, H. (2004). Medical geography: Critical medical and health geography? *Progress in Human Geography, 28*(2), 246–257.

Parr, H. (2008). *Mental health and social space: Towards inclusionary geographies?* John Wiley & Sons.

Parr, H., & Davidson, J., (2010). Mental and emotional health. In Brown, T., McLafferty, S., & Moon, G. (Eds.), *A companion to health and medical geography.* John Wiley & Sons.

Patchett, M. M. (2010). *Putting animals on display: Geographies of taxidermy practice* (Doctoral dissertation). University of Glasgow.

Paterson, M. (2005). Affecting touch: Towards a felt phenomenology of therapeutic touch. In Davidson, J., Bondi, L., & Smith, M. (Eds.), *Emotional geographies.* Ashgate, 161–173.

Paul, B. K. (1994). Commentary on Kearn's 'Place and health: Toward a reformed medical geography.' *The Professional Geographer, 46*(4), 504–505.

Pavlidis, A., & Fullagar, S. (2013). Becoming roller derby grrrls: Exploring the gendered play of affect in mediated sport cultures. *International Review for the Sociology of Sport, 48*(6), 673–688.

Percy-Smith, B., & Matthews, H. (2001). Tyrannical spaces: Young people, bullying and urban neighbourhoods. *Local Environment, 6*(1), 49–63.

Peter, E. (2002). The history of nursing in the home: Revealing the significance of place in the expression of moral agency. *Nursing Inquiry, 9*(2), 65–72.

Phillips, D. R. (1990). *Health and health care in the Third World.* Longman Scientific and Technical.

Phillips, D. R. (2002). *Ageing in the Asia-Pacific region: Issues, policies and future trends.* Routledge.

Phillips, D. R., & Verhasselt, Y. (1994). Health and development: Retrospect and prospect. In Phillips, D. R., & Verhasselt, Y. (Eds.), *Health and development.* Routledge, 301–318.

Phillips, D. R., & Vincent, J. A. (1986). Private residential accommodation for the elderly: Geographical aspects of developments in Devon. *Transactions of the Institute of British Geographers,* 155–173.

Philo, C. (2007). A vitally human medical geography? Introducing Georges Canguilhem to geographers. *New Zealand Geographer, 63*(2), 82–96.

Philo, C. (2012). A 'new Foucault' with lively implications – or 'the crawfish advances sideways'. *Transactions of the Institute of British Geographers, 37*(4), 496–514.

Philo, C., Cadman, L., & Lea, J. (2015). New energy geographies: A case study of yoga, meditation and healthfulness. *Journal of Medical Humanities, 36*(1), 35–46.

Pile, S. (2010). Emotions and affect in recent human geography. *Transactions of the Institute of British Geographers, 35*(1), 5–20.

Pink, S. (2008). Sense and sustainability: The case of the Slow City movement. *Local Environment, 13*(2), 95–106.

Pitt, H. (2014). Therapeutic experiences of community gardens: Putting flow in its place. *Health & Place, 27,* 84–91.

Pouliou, T., & Elliott, S. J. (2010). Individual and socio-environmental determinants of overweight and obesity in urban Canada. *Health & Place, 16,* 389–398.

Pred, R. (2005). *Onflow: Dynamics of consciousness and experience.* Bradford Books.

Radcliffe, P. (1999). Geographical mobility, children and career progress in British professional nursing. *Journal of Advanced Nursing, 30*(3), 758–768.

Rapport, F., Doel, M. A., Greaves, D., & Elwyn, G. (2006). From manila to monitor: Biographies of general practitioner workspaces. *Health, 10*(2), 233–251.

Rapport, F., Doel, M. A., & Elwyn, G. (2007). Snapshots and snippets: General practitioners' reflections on professional space. *Health & Place, 13*(2), 532–544.

Ravn, S., & Duff, C. (2015). Putting the party down on paper: A novel method for mapping youth drug use in private settings. *Health & Place, 31,* 124–132.

Relph, E. (1976). *Place and placelessness.* Pion.

Rican, S., & Salem, G. (2010). Mapping disease. In Brown, T., McLafferty, S., & Moon, G. (Eds.), *A companion to health and medical geography.* John Wiley & Sons.

Rich, E. (2010). Obesity assemblages and surveillance in schools. *International Journal of Qualitative Studies in Education, 23*(7), 803–821.

Richmond, C. (2016). Applying decolonizing methodologies in environment-health research: A community-based film project with Anishinabe communities. In Fenton, N. E., & Baxter, J. (Eds.), *Practicing qualitative methods in health geographies.* Routledge.

Riva, M., & Curtis, S. (2012). The significance of material and social contexts for health and wellbeing in rural England. In Atkinson, S., Fuller, S., & Painter, J. (Eds.), *Wellbeing and place*. Ashgate.

Roe, E., & Greenhough, B. (2014). Experimental partnering: Interpreting improvisatory habits in the research field. *International Journal of Social Research Methodology, 17*(1), 45–57.

Rogers, D. J., Randolph, S. E. (2000). The global spread of malaria in a future, warmer world. *Science, 289*(5485), 1763–1766.

Rose, E. (2012). Encountering place: A psychoanalytic approach for understanding how therapeutic landscapes benefit health and wellbeing. *Health & Place, 18*(6), 1381–1387.

Rosenberg, M. W., Moore, E. G. (1997). The health of Canada's elderly population: Current status and future implications. *Canadian Medical Association Journal, 157*(8), 1025–1032.

Ross, S. J., Polsky, D., & Sochalski, J. (2005). Nursing shortages and international nurse migration. *International Nursing Review, 52*(4), 253–262.

Rowles, G. D. (2000). Habituation and being in place. *OTJR: Occupation, Participation and Health, 20*(1), 52S–67S.

Ruddick, S. (2012). Power and the problem of composition. *Dialogues in Human Geography, 2*(2), 207–211.

Sabel, C., Pringle, D., & Schaestrom, A. (2010). Infectious disease diffusion. In Brown, T., McLafferty, S., & Moon, G. *A companion to health and medical geography*. John Wiley & Sons.

Sampson, R., & Gifford, S. M. (2010). Place-making, settlement and well-being: The therapeutic landscapes of recently arrived youth with refugee backgrounds. *Health & Place, 16*(1), 116–131.

Saville, S. J. (2008). Playing with fear: Parkour and the mobility of emotion. *Social & Cultural Geography, 9*(8), 891–914.

Schaaf, R. (2012). Place matters: Aspirations and experiences of wellbeing in Northeast Thailand. In Atkinson, S., Fuller, S., & Painter, J. (Eds.), *Wellbeing and place*. Ashgate.

Schuurman, N., Bell, N. J., L'Heureux, R., & Hameed, S. M. (2009). Modelling optimal location for pre-hospital helicopter emergency medical services. *BMC Emergency Medicine, 9*(1), 6.

Schwanen, T., & Atkinson, S. (2015). Geographies of wellbeing: An introduction. *The Geographical Journal, 181*(2), 98–101.

Schwanen, T., & Páez A. (2010). The mobility of older people – an introduction. *Journal of Transport Geography, 18*(5), 591–595.

Schwanen, T., & Wang, D. (2014). Well-being, context, and everyday activities in space and time. *Annals of the Association of American Geographers, 104*(4), 833–851.

Schwanen, T., & Ziegler, F. (2011). Wellbeing, independence and mobility: An introduction. *Ageing and Society, 31*(5), 719–733.

Schwanen, T., Hardill, I., & Lucas, S., (2012). Spatialities of ageing: The co-construction and co-evolution of old age and space. *Geoforum, 43*(6), 1291–1295.

Scott, K. (2015). Happiness on your doorstep: Disputing the boundaries of wellbeing and localism. *The Geographical Journal, 181*, 129–137.

Seamon, D. (1979). *A geography of the lifeworld: Movement, rest and encounter*. Croom Helm.

Segrott, J., & Doel, M. A. (2004). Disturbing geography: Obsessive-compulsive disorder as spatial practice. *Social & Cultural Geography, 5*(4), 597–614.

Shannon, G. W., & Pyle, G. F. (1989). The origin and diffusion of AIDS: A view from medical geography. *Annals of the Association of American Geographers, 79*(1), 1–24.

Sheller, M., & Urry, J. (2006). The new mobilities paradigm. *Environment and planning A, 38*(2), 207–226.

Shields, R. (1997). Flow as a new paradigm. *Space and Culture, 1*(1), 1–7.

Simpson, P. (2010). *Ecologies of street performance: Bodies, affects, politics* (Unpublished doctoral dissertation). University of Bristol.

Simpson, P. (2014). A soundtrack to the everyday: Street music and the production of convivial, healthy public spaces. In Andrews, G. J., Kingsbury, P., & Kearns, R. A. (Eds.), *Soundscapes of and wellbeing in popular music*. Routledge.

Simpson, P (2015). Non-representational theory. Retrieved from www.oxfordbibliographies.com/view/document/obo-9780199874002/obo-9780199874002-0117.xml.

Simpson, P. (2017). A sense of the cycling environment: Felt experiences of infrastructure and atmospheres. *Environment and Planning A, 49*(2), 426–447.

Skinner, M. W., Joseph, A. E. (2011). Placing voluntarism within evolving spaces of care in ageing rural communities. *GeoJournal, 76*(2), 151–162.

Skinner, E., & Masuda, J. R. (2014). Mapping the geography of health inequity through participatory hip hop. In Andrews, G., Kingsbury, P., & Kearns, R. (Eds.), *Soundscapes of wellbeing in popular music*. Routledge.

Smallman-Raynor, M. R., Cliff, A. D. (1991). Civil war and the spread of AIDS in Central Africa. *Epidemiology and Infection, 107*(1), 69–80.

Smallman-Raynor, M., Phillips, D. R. (1999). Late stages of epidemiological transition: Health status in the developed world. *Health & Place, 5*(3), 209–222.

Smith, M., & Davidson, J. (2006). It makes my skin crawl…': The embodiment of disgust in phobias of nature. *Body & Society, 12*(1), 43–67.

Smith, S. J., Pain, R. (2010). Introduction: Geographies of wellbeing. In Smith, S. J., Pain, R., Jones, J.-P., & Marston, S. (Eds.), *Handbook of social geography*. Sage.

Smith, K., Luginaah, I., & Lockridge, A. (2010). 'Contaminated' therapeutic landscape: The case of the Aamjiwnaang First Nation in Ontario, Canada. *Geography Research Forum, 30*, 83–102.

Smyth, F. (2005). Medical geography: Therapeutic places, spaces and networks. *Progress in Human Geography, 29*(4), 488–495.

Snyder, J., Crooks, V. A., Adams, K., Kingsbury, P., & Johnston, R. (2011). The 'patient's physician one-step removed': The evolving roles of medical tourism facilitators. *Journal of Medical Ethics*, 37(9), 530–534.

Soffer, A. K. B. (2015). Replacing and representing patients: Professional feelings and plastic body replicas in nursing education. *Emotion, Space and Society, 16*(1), 11–18.

Solberg, S. M., & Way. C. (2007). The importance of geography and health in nursing research. *Canadian Journal of Nursing Research, 39*(3), 13–18.

Solomon, H. (2011). Affective journeys: The emotional structuring of medical tourism in India. *Anthropology and Medicine, 18*(1), 105–118.

Sperling, J. M., & Decker, J. F. (2007). The therapeutic landscapes of the Kaqchikel of San Lucas Toliman, Guatemala. In Williams, A., (Ed.), *Therapeutic landscapes*. Ashgate

Spinney, J. (2006). A place of sense: A kinaesthetic ethnography of cyclists on Mont Ventoux. *Environment and Planning D, 24*,(5), 709.

Spinney, J. (2015). Close encounters? Mobile methods, (post) phenomenology and affect. *Cultural Geographies, 22*(2), 231–246.

Stephens, L., Ruddick, S., & McKeever, P. (2015). Disability and Deleuze: An exploration of becoming and embodiment in children's everyday environments. *Body & Society*, *21*(2), 194–220.

Stock, R. (1983). Distance and the utilization of health facilities in rural Nigeria. *Social Science & Medicine*, *17*(9), 563–570.

Stuart, M., Elliott, J., & Toman, C. (2008). *Place and practice in Canadian nursing history*. UBC Press.

Sullivan, N. (2012). Enacting spaces of inequality: Placing global/state governance within a Tanzanian hospital. *Space and Culture*, *15*(1), 57–67.

Tan, Q. H. (2012). Towards an affective smoking geography. *Geography Compass*, *6*(9), 533–545.

Taussig, M. T. (1993). *Mimesis and alterity: A particular history of the senses*. Psychology Press.

Taylor, S. M., Dear, M. J., & Hall, G. B. (1979). Attitudes toward the mentally ill and reactions to mental health facilities. *Social Science & Medicine. Part D: Medical Geography*, *13*(4), 281–290.

Thien, D. (2005). After or beyond feeling? A consideration of affect and emotion in geography. *Area*, *37*: 450–456.

Thomas, B. S. P. (2013). The global phenomenon of urbanization and its effects on mental health. *Issues in Mental Health Nursing*, *34*(3), 139–140.

Thompson, L., Pearce, J., & Barnett, J. R. (2007). Moralising geographies: Stigma, smoking islands and responsible subjects. *Area*, *39*(4), 508–517.

Thorpe, H., & Rinehart, R. (2010). Alternative sport and affect: Non-representational theory examined. *Sport in Society*, *13*,(7–8), 1268–1291.

Thrift, N. (1996). *Spatial formations*. Sage.

Thrift, N. (1997). The still point: Resistance, expressive embodiment and dance. In Pile, S., & Keith, M. (Eds.), *Geographies of resistance*. Routledge, 124–151.

Thrift, N. (1999). Steps to an ecology of place. In Massey, D., Allen, J., & Sarre P. (Eds.), *Human geography today*. Polity Press, 295–322.

Thrift, N. (2000). Afterwords. *Environment and Planning D: Society & Space*, *18*(2), 213–255.

Thrift, N. (2003). Space: The fundamental stuff of geography. In Holloway, S., Rice, S., & Valentine, G. (Eds.), *Key concepts in geography*. Sage, 85–96.

Thrift, N. (2004a). Summoning life. In Goodwin, M., Crang, P., & Cloke, P. (Eds.), *Envisioning human geographies*. Routledge, 81–103.

Thrift, N. (2004b). Intensities of feeling: Towards a spatial politics of affect. *Geografiska Annaler B*, *86*, 57–78.

Thrift, N. (2004c). Movement-space: The changing domain of thinking resulting from the development of new kinds of spatial awareness. *Economy and Society*, *33*(4), 582–604.

Thrift, N. (2008). *Non-representational theory: Space, politics, affect*. Routledge.

Thrift, N. (2011). Lifeworld Inc – and what to do about it. *Environment and Planning D: Society and Space*, *29*(1), 5–26.

Thrift, N., & Dewsbury, J. D. (2000). Dead geographies – and how to make them live. *Environment and Planning D: Society and Space*, *18*(4), 411–432.

Timmons, S., Crosbie, B., & Harrison-Paul, R. (2010). Displacement of death in public space by lay people using the automated external defibrillator. *Health & Place*, *16*(2), 365–370.

Tolia-Kelly, D. (2006). Affect – an ethnocentric encounter? Exploring the 'universalist' imperative of emotional / affectual geographies. *Area*, *38*(2), 213–217.

Tonnellier, F., & Curtis, S. (2005). Medicine, landscapes, symbols: *The country doctor by Honoré de Balzac*. *Health & Place*, *11*(4), 313–321.

Townley, G., Kloos, B., & Wright, P. A. (2009). Understanding the experience of place: Expanding methods to conceptualize and measure community integration of persons with serious mental illness. *Health & Place*, *15*(2), 520–531.

Townshend, T. & Lake, A. A. (2009). Obesogenic urban form: Theory, policy and practice. *Health & Place*, *15*, 909–916.

Tuan, Y. F. (1977). *Space and place: The perspective of experience*. University of Minnesota Press.

Tuan, Y. T. (2004). *Home*. In Harrison, S., Pile, S., & Thrift, N. (Ed.), *Patterned ground: Entanglements of nature and culture*. Reaktion Books, 164–165.

Tucker, I. (2010a). Everyday spaces of mental distress: The spatial habituation of home. *Environment and Planning D: Society and Space*, *28*(3), 526–538.

Tucker, I. (2010b). Mental health service user territories: Enacting 'safe spaces' in the community. *Health 14*(4), 434–448.

Tucker, I. M. (2017). Shifting landscapes of care and distress: A topological understanding of rurality. In Soldatic, K., & Johnson, K. (Eds.), *Disability and rurality: Identity, gender and belonging*. Routledge.

Twigg, L., & Cooper, L. (2010). Healthy behavior. In Brown, T., McLafferty, S., & Moon, G. (Eds.), *A companion to health and medical geography*. Wiley-Blackwell, 460–475.

Vandemark, L. M. (2007). Promoting the sense of self, place, and belonging in displaced persons: The example of homelessness. *Archives of Psychiatric Nursing*, *21*(5), 241–248.

Vannini, P. (2009). Nonrepresentational theory and symbolic interactionism: Shared perspectives and missed articulations. *Symbolic Interaction*, *32*(3), 282–286.

Vannini, P. (Ed.). (2015a). *Non-representational methodologies: Re-envisioning research*. Routledge.

Vannini, P. (2015b). Non-representational research methodologies: An introduction. In Vaninni, P. (Ed.), *Non-representational methodologies: Re-envisioning research*. Routledge.

Vannini, P. (2015c). Enlivening ethnography through the irrealis mood: In search of a more-than-representational style. In Vaninni, P. (Ed.), *Non-representational methodologies: Re-envisioning research*. Routledge.

Vannini, P. (2015d). Non-representational ethnography: New ways of animating lifeworlds. *Cultural Geographies*, *22*(2), 317–327.

Veenstra, G., Luginaah, I., Wakefield, S., Birch, S. Eyles, J., & Elliott, S. (2005). Who you know, where you live: Social capital, neighbourhood and health. *Social Science & Medicine*, *60*(12), 2799–2818.

Vertinsky, P. A., & Bale, J. (2004). *Sites of sport: Space, place, experience*. Routledge.

Waitt, G., & Cook, L. (2007). Leaving nothing but ripples on the water: Performing ecotourism natures. *Social & Cultural Geography*, *8*(4), 535–550.

Waitt, G., & Stanes, E. (2015). Sweating bodies: Men, masculinities, affect, emotion. *Geoforum*, *59*(1), 30–38.

Wakefield, S. E., Elliott, S. J., Cole, D. C., & Eyles, J. D. (2001). Environmental risk and (re) action: Air quality, health, and civic involvement in an urban industrial neighbourhood. *Health & Place*, *7*(3), 163–177.

Walton, E. (2014). Vital places: Facilitators of behavioral and social health mechanisms in low-income neighborhoods. *Social Science & Medicine, 122*, 1–12.

Ward Thompson, C., & Aspinall, P. A. (2011). Natural environments and their impact on activity, health, and quality of life. *Applied Psychology: Health and Well-Being, 3*(3), 230–260.

Weinstein Agrawal, A., Schlossberg, M., & Irvin, K. (2008). How far, by which route and why? A spatial analysis of pedestrian preference. *Journal of Urban Design, 13*(1), 81–98.

Wendt, D. C., & Gone, J. P. (2012). Urban-indigenous therapeutic landscapes: A case study of an urban American Indian health organization. *Health & Place, 18*(5), 1025–1033.

West, E., & Barron, D. N. (2005). Social and geographical boundaries around senior nurse and physician leaders: An application of social network analysis. *Canadian Journal of Nursing Research, 37*(3), 132–149.

Wheeler, A. (2012). Am I an eco-warrior now? Place, wellbeing and the pedagogies of connection. In Atkinson, S., Fuller, S., & Painter, J. (Eds.), *Wellbeing and place.* Ashgate.

Whitelegg, J. (1987). A geography of road traffic accidents. *Transactions of the Institute of British Geographers, 12*(2), 161–176.

Widener, M. J., & Hatzopoulou, M. (2016). Contextualizing research on transportation and health: A systems perspective. *Journal of Transport & Health, 3*(3), 232–239.

Wiles, J. (2003). Daily geographies of caregivers: Mobility, routine, scale. *Social Science & Medicine, 57*(7), 1307–1325.

Wiles, J. (2011). Reflections on being a recipient of care: Vexing the concept of vulnerability. *Social & Cultural Geography, 12*(6), 573–588.

Wiles, J., Allen, R. E., Palmer, A. J., Hayman, K. J., Keeling, S., & Kerse, N. (2009). Older people and their social spaces: A study of well-being and attachment to place in Aotearoa New Zealand. *Social Science & Medicine, 68*(4), 664–671.

Williams, A. (2002). Changing geographies of care: Employing the concept of therapeutic landscapes as a framework in examining home space. *Social Science & Medicine, 55*(1), 141–154.

Williams, A. M. (2006). Restructuring home care in the 1990s: Geographical differentiation in Ontario, Canada. *Health & Place, 12*(2), 222–238.

Williams, A. (2007). Healing landscapes in the Alps: Heidi by Johanna Spyri. In Williams, A. (Ed.), *Therapeutic landscapes.* Ashgate.

Williams, A. (2010). Spiritual therapeutic landscapes and healing: A case study of St. Anne de Beaupre, Quebec, Canada. *Social Science & Medicine, 70*(10), 1633–1640.

Williams, A. M, King, R., Warnes, A, & Patterson, G. (2000). Tourism and international retirement migration: New forms of an old relationship in southern Europe. *Tourism Geographies, 2*(1), 28–49.

Willis, A. (2009). Restorying the self, restoring place: Healing through grief in everyday places. *Emotion, Space and Society, 2*(2), 86–91.

Wilson, K. (2003). Therapeutic landscapes and First Nations peoples: An exploration of culture, health and place. *Health & Place, 9*(2), 83–93.

Wilton, R., & DeVerteuil, G. (2006). Spaces of sobriety/sites of power: Examining social model alcohol recovery programs as therapeutic landscapes. *Social Science & Medicine, 63*(3), 649–661.

Wilton, R., DeVerteuil, G., & Evans, J. (2014). 'No more of this macho bullshit': Drug treatment, place and the remaking of masculinity. *Transactions: Institute of British Geographers*, *39*(2), 291–303.

Witten, K., Hiscock, R., Pearce, J., Blakely, T. (2008). Neighbourhood access to open spaces and the physical activity of residents: A national study. *Preventive Medicine*, *47*(3), 299–303.

Wood, V. J., Curtis, S. E., Gesler, W., Spencer, I. H., Close, H. J., Mason, J. M., & Reilly, J. G. (2013). Spaces for smoking in a psychiatric hospital: Social capital, resistance to control, and significance for 'therapeutic landscapes'. *Social Science & Medicine*, *97*(1), 104–111.

Wylie, J. (2005). A single day's walking: Narrating self and landscape on the South West Coast Path. *Transactions of the Institute of British Geographers*, *30*,(2), 234–247.

Wylie, J. (2009). Landscape, absence and the geographies of love. *Transactions of the Institute of British Geographers*, *34*(3), 275–289.

Yanos, P. T. (2007). Beyond 'Landscapes of Despair': The need for new research on the urban environment, sprawl, and the community integration of persons with severe mental illness. *Health & Place*, *13*(3), 672–676.

Zenk, S. N., Schulz, A. J., Matthews, S. A., Odoms-Young, A., Wilbur, J., Wegrzyn, L., ... Stokes, C. (2011). Activity space environment and dietary and physical activity behaviors: A pilot study. *Health & Place*, *17*(5), 1150–1161.

Index

Printed and bound by CPI Group (UK) Ltd, Croydon, CR0 4YY

24/10/2024

01778282-0007